LEADERSHIP AND COMMAND

This book must be returned after 30 days.
Replacement cost will be invoiced if book is not returned.

Vous devez retourner ce livre après 30 jours.
Le coût de remplacement sera facturé si le
livre n'est pas retourné.

RETURN BY/DATE DE RETOUR		

The Operational Art:
Canadian Perspectives

Leadership and Command

Edited by
Allan English

CANADIAN DEFENCE ACADEMY PRESS

Canadian Defence Academy Press
PO Box 17000 Stn Forces
Kingston, Ontario   K7K 7B4

Produced for the Canadian Defence Academy Press
by 17 Wing Winnipeg Publishing Office.
WPO30172

Cover Photo: Silvia Pecota

Library and Archives Canada Cataloguing in Publication

The operational art : Canadian perspectives : leadership and command /
edited by Allan English.

Issued by Canadian Defence Academy.
Includes bibliographical references.

Soft Cover
ISBN 0-662-43220-7
Cat. no.: D4-3/2-2006E

Hard Cover
ISBN 0-662-43272-X
Cat. No.: D4-3/2-2006-1E

1. Military art and science--Canada. 2. Canada--Armed Forces.
3. Canada--Armed Forces--Officers. 4. Operational art (Military science).
5. Command of troops. I. English, Allan D. (Allan Douglas), 1949- II. Canadian
Defence Academy. III. Title: Leadership and command.

UA600.O63 2006 355.30971 C2006-980116-9

Printed in Canada.
  3 5 7 9 10 8 6 4 2

TABLE OF CONTENTS

FOREWORD

The Canadian Defence Academy (CDA) was created in 2002 to champion, govern and manage professional-development reform initiatives in the Canadian Forces (CF). The CDA is also the body that will institutionalize and maintain the momentum behind these reforms, which contribute to the CF's professional-development strategic objectives. Key among these objectives is fostering intellectual development and critical thinking within Canada's military, and the transformation of the CF into a learning organization. "Professional development," explained General Raymond Henault, the former Chief of the Defence Staff, "is at the heart of the profession of arms. The Canadian Defence Academy will play a vital role in the reform and transformation of our professional standards and competencies."

The publication of this book is an initiative under the strategic leadership writing project of the Canadian Forces Leadership Institute and should stimulate debate about the profession of arms in Canada. Readers are invited to join the debate and make their own contribution to military professionalism in this country.

Major-General Paul Hussey
Commander, Canadian Defence Academy

PREFACE

The Operational Art: Canadian Perspectives — Leadership and Command was commissioned by the Canadian Forces Leadership Institute as part of its strategic leadership writing project, which is designed to compile a body of knowledge relating to Canadian military leadership (that is, past, present and future) that can be used by Canadian Forces educational and training institutions to generate effective military leaders, as well as educate the public. It is our intent that Canadian officers have the opportunity to learn from Canadian examples. After all, although our military culture and experience, for instance, is different from that of our close allies the Americans and the British, it is equally rich and distinctive.

In this vein, *The Operational Art: Canadian Perspectives — Leadership and Command* is a seminal work that comes on the heels of and complements the Canadian Forces College's *The Operational Art: Canadian Perspectives — Context and Concepts*. It examines arguably the most essential component of the operational art, namely, leadership and command. Importantly, its chapters are written by a wide array of authors who span the gamut from practitioner to theoretician. In the end, these diverse individuals with both practical experience and theoretical knowledge bring insight and depth to the discussion.

I believe you will find *The Operational Art: Canadian Perspectives — Leadership and Command* to be a stimulating addition to the Canadian Defence Academy Press collection. It deals with such dynamic topics as the differing leadership styles in the three distinctive environments, the complex operating environment, the evolution of combined and joint command structures, as well as leadership and command on expeditionary operations. This volume is sure to expand your knowledge and fuel the Canadian discourse on the operational art.

Colonel Bernd Horn
Director, Canadian Forces Leadership Institute

INTRODUCTION

Allan English

Commissioned by the Canadian Forces Leadership Institute, *The Operational Art: Canadian Perspectives — Leadership and Command* was written to complement *The Operational Art: Canadian Perspectives — Context and Concepts*, which was published by the Canadian Defence Academy Press in 2005 and based on work done by staff and students at the Canadian Forces College (CFC) in Toronto. Its purpose was to offer perspectives on distinct Canadian approaches to the operational art, based on our national and military culture and historical experience.

Leadership and Command focuses on specific aspects, and arguably the most important aspects, of operational art: leadership and command. Given the nature of the topic, the contributions come not only from papers written by staff or students at CFC, but also from a more diverse group of authors. Therefore, this volume includes contributions by a serving Chief of the Defence Staff (CDS), General Rick Hillier, and a former CDS who is now Chairman of the Military Committee of the North Atlantic Treaty Organization, General Ray Henault. Other military professionals, some of whom wrote their contributions while staff or students at CFC, bring further practitioners' insights to leadership and command and the operational art. These practitioners' insights are complemented by a new offering by two of Canada's leading theoreticians of command, Ross Pigeau and Carol McCann from Defence Research and Development Canada, Toronto. The field of contributors is rounded out by Canadian academics whose fields of expertise are related to the subject of this book.

In Chapter One, I begin by examining differences in leadership among the army, navy, and air force, based on differences in national and service cultures. I conclude that even in the unified Canadian Forces (CF), where a significant amount of training and education is conducted in a joint environment, leaders have different "masks of command" based on their experience as junior officers in the Canadian Army, Navy or Air Force. These different masks of command can be distinguished by, among other things, varying proportions of "heroic" and "technical" styles of leadership.

Chapters Two to Four deal with theoretical aspects of leadership, command and the operational art. In Chapter Two, Richard Gimblett examines naval command and leadership, focussing on command styles found in the Canadian Navy. He argues that the nature of operations at sea defines many aspects of naval command and that national cultures and each navy's experience also shape naval command styles. Gimblett concludes that the Canadian Navy, based on its culture of professionalism and its experience, is well placed to lead coalition and alliance naval operations.

The third chapter, by Christian Rousseau, examines decision making in the complex and dynamic systems that characterize the modern battlespace. He concludes that modern practitioners of the operational art can perform better in the chaos created by those systems if their selection and development have been rigorous and have addressed such issues as identifying their capacity to tolerate chaos and providing them with experience that engenders expertise. Rousseau notes that the deliberate planning exercises used at many staff colleges do not provide adequate preparation for modern practitioners of the operational art.

Chapter Four, by Ross Pigeau and Carol McCann, continues their theoretical work on military command and control by exploring the concept of intent. They argue that for those militaries that have command philosophies based on mission command, common intent is the primary means for achieving co-ordinated action, especially if those militaries also embrace concepts such as effects-based operations and network-enabled operations. Besides joint operations involving primarily military forces, the concept of common intent also has implications for "integrated" operations involving government agencies, non-government agencies, and military forces.

The next three chapters examine issues related to the creation and evolution of joint and combined command structures in Canada during the post–Cold War era. Chapter Five by Chris Weicker describes the evolution of the Canadian Forces' operational-level command and control structure over the past decade, based on documentary sources. He argues that a product of that evolution, the Joint Operations Group (JOG), should have been given the responsibility for the operational-level command and control of contingency operations, and he presents two options that could have been used to implement his proposal. Even though the Joint Operations Group was disbanded during the current CF reorganization based on transformation initiatives, its functions will be assumed by other organizations. Therefore, it is important to understand the context behind the evolution of CF operational-level command and control arrangements such as the JOG.

Chapter Six is based on a personal account of the evolution of the Canadian Forces' operational-level command and control structure over the past decade, by Ray Henault, a former Deputy Chief of the Defence Staff and CDS. His account brings new insights into that evolution, which he significantly influenced during his eight years in National Defence Headquarters at the end of the twentieth century and the beginning of the twenty-first century.

Chapter Seven by Kenneth Hansen examines Canada's "medium-power dilemma" from a largely naval command and control perspective. The dilemma occurs when national naval capability, optimized for domestic sovereignty, is not always adequate for the high-end combat capability and endurance required by "blue water" alliance naval tasks. Hansen argues that adequate resources must be devoted to develop appropriate Canadian doctrine and to provide adequate numbers of experienced staff officers to carry out alliance naval and joint tasks. Otherwise, without the doctrine and staffs to guide procurement and force-employment decisions, Canada may not have forces that are optimized to meet its needs and will be compelled to rely on foreign sources for doctrinal guidance and essential military staff skills in order to create its forces and conduct its operations.

The next two chapters discuss issues of leadership and command in Canadian expeditionary operations. Chapter Eight by Howard Coombs and Rick Hillier is a case study of the Canadian-led International Security Assistance Force (ISAF) Rotation V in Afghanistan during 2004. This study demonstrates that, unlike the technology-focused networks in the network-centric warfare model, Canadian network-enabled operations during peace support operations have used networks that are not solely reliant on technology but are hybrid, consisting of a mixture of information and social networks. These human-centric networks are used to formulate military plans that feature unified and balanced efforts by all agencies to achieve shared intent and collectively promote the conditions necessary for success. Recent Canadian experiences of peace support operations may provide the basis for a new doctrine that includes the formal utilization of network-enabled operations during interventions in post-conflict environments.

Chapter Nine by Daniel Gosselin examines decision making for Canadian expeditionary operations and its impact on mission command in the CF. He argues that the increased compression of the levels of war in recent times has centralized CF decision making and decreased the use of mission command in CF expeditionary operations. He concludes that, in devising a CF command philosophy for the future, the CF must create a Canadian command framework developed for Canadian national requirements, using contemporary command principles. If a mission-command philosophy is to be adopted for CF expeditionary operations, the management of risk must be the key factor influencing the development of control structures and processes.

As this introduction was being written, I received a discouraging e-mail that summarized a recent discussion between two Canadian senior officers. I found the e-mail to be discouraging because it reflected ideas, still persisting among some CF senior officers, which run counter to the principles of Canadian Professional Military Education (PME). These principles hold that education is one of the pillars of the profession of arms in Canada. Those who hold contrary views question the relevance of academic rigour in CF PME, fearing that by taking academically rigorous courses they will be turned into "academics" or "theoreticians" and will thereby no longer be effective military officers. The e-mail concluded by saying that a senior officer had no interest in becoming a "theorist," but that he did have a desire to improve his ability as an officer.

Given that this book and its predecessor, *The Operational Art: Canadian Perspectives — Context and Concepts*, are composed of contributions from senior officers who have written with academic rigour and who have presented theoretical concepts related to the operational art, it is worth pointing out here the value of such work to Canadian officers. As members of the profession of arms in Canada, senior officers are required to pursue "the highest standards of the required expertise" for their profession.[1] As noted by the CF's profession of arms manual, *Duty with Honour*:

The expertise required by the military professional is determined by the direction, operation and control of a human organization whose primary function is the application of military force. Such an organization is supported by a sophisticated body of theoretical and practical knowledge and skills that differ from those in any other profession.

The foundation for this expertise resides in a deep and comprehensive understanding of the theory and practice of armed conflict — a theory that incorporates the history of armed conflict and the concepts and doctrine underpinning the levels inherent in the structure of conflict, ranging from the tactical and operational to the military strategic and political-military (policy) levels. Increasingly, the military professional, especially when advancing in rank, must master the domain of joint, combined and inter-agency operations and, in the highest ranks, have an expert understanding of national security issues. An understanding of how the law, both national and international, regulates armed conflict is also very important.[2]

A vital way of imparting the expertise required to master the profession of arms is through PME. As professional education, PME courses contain both theory and practice, and a great deal of the theory in these courses supports the practice of the profession. For example, just as professional engineers must master certain theories founded in the physical sciences to practise their profession, military professionals must master theories of war, leadership and command to be competent to practise their profession. The excuse given by some that they are too busy doing operations to engage in serious professional military education seems a rather strange argument to many in other professions, such as medicine or law, who accept that they must set time aside to upgrade themselves professionally on a regular basis.

My disappointment in receiving the e-mail referred to above was largely based on the fact that some senior officers in the CF still seem to think that theoretical knowledge and academic rigour are incompatible with the duties of military officers. The US military has recognized for years that academic rigour is essential to PME and that theory is not taught for theory's sake or to make military officers theorists, but to enable them to apply relevant theories to the practice of their profession. Therefore, one of the aims of this book, and its predecessor, is to provide theory and accounts of experience that will in some way help military officers practise their vocation in a professional manner. Another aim is to encourage other military professionals in this country to contribute to the growing body of knowledge supporting the practice of the profession of arms in Canada. Finally, I hope that *Leadership and Command* will help to put to rest the myth that theoretical knowledge and academic rigour are incompatible with the duties of military officers.

NOTES

1   Canada, Department of National Defence, *Duty with Honour* (Kingston, ON: Canadian Defence Academy, 2003), 11.

2   Ibid., 17.

CHAPTER 1

THE MASKS OF COMMAND: LEADERSHIP DIFFERENCES IN THE CANADIAN ARMY, NAVY AND AIR FORCE[1]

Allan English

At the end of the twentieth century, warfare was increasingly characterized by operations where the forces of different nations fought together in coalitions, and different services (army, navy and air force) worked together closely to accomplish a mission. These operations are often called *combined*[2] and *joint*,[3] respectively. At the beginning of the twenty-first century, new security challenges have caused many Western nations to have their armed forces work much more closely with other agencies, and this phenomenon has added expressions like *Joint, Interagency, Multinational, and Public* (JIMP), 3D (defence, diplomacy and development), and integrated to the national security lexicon. Working in these environments creates leadership challenges at all rank levels in the military. While there is some literature on the challenges of working in multi-national coalitions, the literature on leadership in joint operations, let alone in the new integrated operating environment, is extremely sparse despite the fact that joint operations are even more numerous than combined operations and that integrated operations are becoming the norm.

Some may assume that the Canadian Forces (CF) has overcome the problem of Environmental (or what most nations refer to as service)[4] differences in leadership because it is, in law, a unified service. Yet, even in the unified CF, where basic officer training and many courses are conducted in a joint environment, leaders spend their most formative years in a single service culture that shapes their attitudes, values and beliefs about what is an appropriate leadership style. These differences have been recognized in recent CF doctrinal manuals.

Two Canadian Forces publications have recently codified and described in detail, for the first time, what it means to be a leader in the Canadian Forces. As well as providing doctrinal guidance for members of the CF, *Duty with Honour* and *Leadership in the CF: Conceptual Foundations* (hereafter, *Conceptual Foundations*) provide frameworks and theoretical models for analyzing Canadian military leadership. Both these publications acknowledge that despite many similarities, there are Environmental differences in culture[5] based on the unique physical environments in which the Canadian Army, Navy and Air Force operate. These operating environments have produced a unique body of professional knowledge,[6] experience and, therefore, culture for each Environment.

Duty with Honour acknowledges that differences among the three Environments are "essential for readiness, generating force and sustaining a multi-purpose, combat-capable force."[7] And because of these differences "all three Environments often manifest certain elements of the [CF's] ethos in different ways; for example, the influence of history, heritage and tradition, or how team spirit is promoted and manifested."[8] Consequently, *Duty with Honour* recognizes that the CF's "ethos must accommodate the separate identities forged by combat at sea, on land and in the air."[9] *Conceptual Foundations* notes that "leaders are formed and conditioned by their social environment and culture";[10] therefore, we can expect to see differences in leadership styles in the Canadian Army, Navy and Air Force based on these Environmental differences in professional expertise and culture. These Environmental differences also influence judgments about what constitutes "good" and "bad" leadership styles.

One of the biggest problems in the CF today is a lack of understanding about the differences in Environmental leadership styles. For example, in conversations with the author, some army officers have characterized certain senior leaders from the other services in joint appointments as indecisive or not forceful enough, and some army officers have even remarked on the lack of physical fitness or the small stature of air force and navy leaders in the context of their less than adequate leadership. On the other hand, some officers of the other two services have from time to time described certain senior army leaders in joint appointments as "all muscle and no brains" because they put physical fitness ahead of intellectual competency, or as "micromanagers" because they try to make forceful interventions in areas where they have little expertise.

Many of these views are predicated on service-based expectations about what good leadership looks like. Some of the views are based on stereotypes, and others on fact. However, we currently have very little in the way of research on this topic to sort myth from reality. In fact we have not even identified in any systematic way all the service-based views on leadership. Many of these views have historical roots; therefore, the approach taken in this chapter will be to put the leadership differences of the Canadian Army, Navy and Air Force in a historical context by looking at aspects of how and why they developed in the ways they did develop, and then by speculating on how they have been evolving. Perhaps by examining the question of inter-service leadership over a relatively long period of time we can come to a better understanding of the challenges of leadership in joint and integrated environments in the twenty-first century. The chapter will conclude with suggestions for future research in this area and with the implications that the service leadership differences might have for joint force commanders, particularly at the operational level. *Duty with Honour* notes that in this country leadership theory and concepts related to the profession of arms are still evolving.[11] Therefore, the aim of this chapter is to make a contribution to the evolving theories of military leadership and professional concepts by examining service (or Environmental) differences in leadership in a Canadian historical context.

While most of the military leadership literature focuses on the experience of land forces, almost all military personnel know from their own experience that there are distinct differences in the leadership styles commonly used in the army, navy, and air force. Each service has a unique culture that influences acceptable leadership styles in that service. At the same time, each nation has a culture that is another variable in the leadership equation; therefore, studies done by other nations are not necessarily applicable to the Canadian context. To address some of the gaps in the literature I will examine leadership differences in the context of the cultures of the Canadian services. I have emphasized air force and navy leadership experiences here in an attempt to widen the field of leadership studies beyond the existing land-centric focus. I am also focusing mainly on officer leadership in the three services because leadership by a non-commissioned officer (NCO) (or non-commissioned member [NCM], in CF parlance) could be a separate chapter on its own. Furthermore, I accept that personalities can have a greater impact on leadership style than can service background, but I will leave that field to others to examine. Although there are many similarities in service leadership styles, my emphasis will be on the more neglected, yet I believe equally important, aspect of differences in leadership as this has become particularly relevant in today's context of joint and integrated operations where leaders of the three services interact more regularly than they did in the past.

Leadership in History

The study of military leadership and its culture is most effective when conducted as a multi-disciplinary endeavour, where each discipline contributes to the endeavour. History's contribution to this undertaking is to provide both data and context. Historians specialize in the evaluation of sources — everything from documents held in archives to oral histories — to produce verifiable data for the study of past leadership. Perhaps just as important, historians describe the times in which military leaders lived, including the culture that shaped the leaders and in which they exercised command. As Sir Basil Liddell Hart put it, history tries "to find out what happened while trying to find out why it happened." In so doing, it seeks causal relationships between events, which can provide analogies that may not teach us exactly what to do today but can teach common mistakes. Liddell Hart also tells us that history has a practical value because historical experience is infinitely longer, wider and more varied than individual experience.[12]

Heroic Leadership. One of the most popular historical books on military leadership, and the one on which I have based my title, is John Keegan's *The Mask of Command*, "a book about the technique and the ethos of leadership and command." Keegan argues that European culture produced a distinctive leadership style that joined Alexander and Wellington across the centuries in "motive and method," despite subtle shifts in culture that made them somewhat different.[13]

While every individual mask of command is unique (based on factors such as personality, previous experience, education and so on), some of the framework of the mask may be common to all three services, especially in Canada where a significant amount of officer leadership education and training is done in a tri-service environment. Nevertheless, since most formative operational leadership experiences occur during an officer's early years in the military, and since much of this time is spent in a single Environment, each officer's mask bears a distinctive service imprint.

A key theme in Keegan's book is that good leaders authenticate themselves in their leadership role by sharing risks with their followers. This cultivates a kinship between leaders and their followers and gives leaders the moral legitimacy, beyond their legal authority, that they must have in order to be successful. Keegan defined the heroic style of leadership as "aggressive, invasive, exemplary, risk-taking."[14] Based on Keegan's analysis, I offer this revised definition of heroic leadership in a twenty-first century context: *conspicuous sharing of risk with subordinates.*

Keegan's examination of leadership was based on a comparison of the masks of command used across the centuries, among various nationalities, but it was primarily focused on land forces. This chapter extends Keegan's analysis by looking at some of the masks of command used in the past 100 years by Canadian leaders in the three environments in which the army, navy and air force fight. However, to properly understand leadership in the past 100 years, I suggest that in addition to Keegan's "heroic" leadership, there is another type of leadership that became increasingly important in the twentieth century and that has become indispensable in the twenty-first century: technical leadership.

Technical Leadership. Technical leadership, as used here, is defined as the ability to influence others to achieve a goal, based on the specialized knowledge or skill of the leader. Technical leadership is exercised either by leaders (for example, pilots) who must be able to actually do the same job as their subordinates do, or by leaders who must have a significant specialized knowledge (for example, naval officers' seamanship skills) of the jobs that their subordinates perform. This type of leadership is critical in the navy and air force where every second at sea or in the air those on board ships and aircraft depend on technology and, by extension, the technical ability (not just the fighting ability) of the crews and their leaders for their very survival. Technical leadership is most clearly different from the traditional concept of army leadership, for example, in pilots who must, as we shall see, be able to demonstrate an acceptable level of flying skill before they will be accepted as leaders.

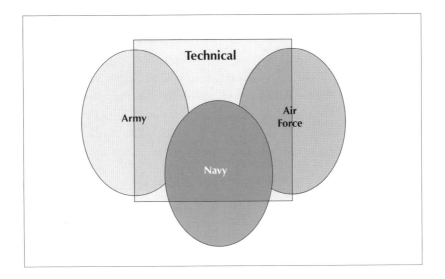

FIGURE 1.1. LEADERSHIP STYLES

While technical leadership is found in all three services in different proportions (as shown in Figure 1.1), the fact that navy and air force leaders are given regular assessments of their technical ability, not just their leadership skills, shows how important the technical aspect of leadership is in these services. This is particularly evident in the air force where aircrew leaders at all levels are given regular check rides by designated standards personnel, who may be their junior in rank.

The land-centric focus of much of the leadership literature leads many, particularly those with little knowledge of military culture, to assume that the masks of command used in the navy and air force are nearly identical to those masks used in the army. This next section examines service differences in leadership in a general context.

Differences in Service Culture. Carl Builder's model of the cultural differences among the American services is a useful starting point because it outlines some general characteristics of Western army, navy and air force cultures today. Builder contends that the touchstone of the US Army's organizational culture is the art of war and the profession of arms; in other words, concepts and doctrine are the glue that unifies the army's separate branches. For the US Navy, the heart of its organizational culture is the navy as an institution, based on tradition, and a maritime strategy, which provide coherence and direction to the navy. The US Air Force, in contrast, he declared, has identified with platforms and air weapons, rooted in a commitment to technical superiority, and it has transformed aircraft or systems into ends in themselves. Builder claimed this lack of an air force vision has had serious repercussions for the force. Writing in the early 1990s, Builder maintained that because the US Air Force had no integrating vision like the US Army's AirLand Battle or the US Navy's Maritime Strategy, it had conceded the intellectual high ground to the other services, particularly the Army.[15] Builder does not discuss the US Marine Corps culture in detail, but it has been described as worshipping "at the altar of its uniqueness," and because of its unique roles it has not been as strongly affected by the end of the Cold War as the other US services have been.[16]

I think we can see some similarities to Canadian service culture in Builder's model; for example, the Canadian Army invests a great deal in doctrine; the Air Force invests very little and remains focused on platforms;[17] and the Navy, with its deep-rooted traditions and maritime strategy *Leadmark*, exhibits many cultural similarities to its American analogue. However, beyond these basic similarities with the American services, Canadian military culture is based on its own historical experience.[18]

In the discussion that follows, I will examine the proposition that Canadian military leadership in the three services is balanced differently between heroic leadership and what I have called technical leadership in unique ways. I will examine this hypothesis about the balance between heroic leadership and technical leadership in the services, focusing somewhat on the perceived cultural dichotomy between the Army and the Air Force to attempt to achieve greater clarity in distinguishing among the subcultures that affect leadership in the Canadian Forces.

Air Force Leadership

This analysis begins with air force leadership for three reasons: (1) air force leadership provides the greatest contrast with the army leadership that is well described in the literature; (2) if the Revolution in Military Affairs (RMA) is leading armed services towards a greater reliance on technology, perhaps the air force style of leadership will be more prevalent in the future; and (3) because the air force is the newest of the three services, we can see some of the roots of its leadership quite clearly.

In this discussion of air force leadership, I will focus on the First and Second World Wars since they form the main basis of our combat experience. Also, since the downsizing of the Directorate of History and Heritage and the cancellation of the post-war volume of the *Official History of the Royal Canadian Air Force*, our knowledge of post–Second World War air force history is quite limited. Also, by necessity, I will focus on aircrew leadership as virtually no research has been published on ground crew leadership.[19]

Until the formation in June 1918 of the Canadian Air Force (which became the Royal Canadian Air Force [RCAF] in April 1924), Canadians who wanted to serve their country in the new dimension of air warfare had to join the British air services, the Royal Flying Corps (RFC) or the Royal Naval Air Service (RNAS) (which combined into the Royal Air Force [RAF] in April 1918). Before having an air service of their own, Canadians made an important contribution to the Imperial flying services. For example, in 1918 about 25 percent of all RAF flying personnel, and perhaps 40 percent of RAF pilots, on the Western front were Canadian, and there were about 22,000 Canadians in the RAF.[20] Therefore, the history of Canadian air force leadership starts with the British air services. Since the RFC had the greatest influence on RAF and RCAF leadership practices, I will focus on it, even though an entirely separate study could be done on naval aviation leadership.

Before the First World War and during the first two years of that war, almost anyone who could get a private pilot's licence and who met basic enrolment standards was accepted to fly for the RFC, which was still a part of the British Army. Pilots held ranks ranging from corporal to general officer, and a pilot's rank was more dependent on his social status than his flying ability. In these early days of military flying, a two-seater aircraft was frequently commanded by the observer, often an artillery officer, who outranked the pilot. This haphazard system of getting aircrew for the RFC was gradually replaced by a formal military training system. In 1918 one of its largest formations (about 20,000 all ranks), the Training Division, was commanded by the highest-ranking Canadian in the RAF, the 28-year-old Brigadier General A.C. Critchley. Interestingly, he was neither a pilot nor an observer but a former cavalry officer who was seconded from the Canadian Corps because of the reputation he had established as an outstanding trainer of land forces. He continued his good work with the Training Division and is credited with modernizing its training methods.[21]

This method of selecting commanders on merit rather than occupation was not uncommon in the British flying services in the First World War. For example, Sir Hugh Trenchard, the "father of the RAF," only learned to fly in 1912 as a major when it seemed that at age 39 his career in the infantry (Royal Scots Fusiliers) had reached a plateau. Three years later he was a major-general commanding the RFC in France. Trenchard had no operational flying experience, let alone combat flying experience; however, this was no barrier to his becoming an effective and highly respected commander of the largest part of the RFC in the field. He personally set the standard for air force leadership, based on the army customs with which he was familiar. His biographer tells us that one morning in 1916, when he was General Officer Commanding of the RFC in France, Trenchard came across an overzealous officer who was punishing some mechanics for infringing a minor regulation, by sending them on a wet cross-country run before breakfast. Trenchard admonished him as follows: "Get this into your thick head.... This is a technical corps.... You're not in the army now, you know."[22] Most of Trenchard's career had been spent in the infantry (in the "golden years" of the British Army's regimental system), and his biographer tells us that "[p]ride in the regiment could never be an abstract sentiment to Trenchard. It had to be felt personally, or nothing."[23] Because Trenchard's remarks were made at least two years before the formation of the RAF as an independent service, this tells us that in the British army at that time there was a recognized form of "technical corps" leadership that was different from that used in the "regular army," or what might be called the combat arms today.

Trenchard and his successors used this style of technical corps leadership to maintain the effectiveness of an organization that suffered heavy losses throughout the war. For example, by 1918, losses among RFC fliers were running as high as 32 percent of unit strength *per month* during offensives.[24] From a leadership perspective this had important consequences. Senior leaders, like Trenchard, tended to be men in their late thirties or older, but because they rarely, if ever, flew in combat, there was little attrition among them. On the other hand, junior leaders, especially at squadron level and below, were being killed at an alarming rate. Aggressive, lead-from-the-front tactics in the air led to high casualties among squadron and flight commanders, and soon squadrons were being routinely led by men in their early twenties. By April 1917 the leadership crisis was so great that squadron commanding officers (COs) were forbidden to fly within five miles of enemy lines. Some returned to fight in the trenches, explaining that they would not risk their subordinates' lives if they could not put their own lives on the line; others broke the rules and flew over enemy territory anyway. It seems things had become so bad by the end of the war that some older army officers, "skilled in the handling of men," were assigned to command squadrons.[25] The rationale for this practice was offered by the official historian of the RFC/RAF:

> A man with a talent for command, who can teach and maintain discipline, encourage his subordinates, and organize the work to be done, will have a good squadron and is free from those insidious temptations which so easily beset commanding officers who have earned distinction as pilots.[26]

We can see by this comment, written just after the war, that some people believed there had been problems with promoting young men in their early twenties to command squadrons, whatever their flying skills might have been. A similar situation arose in the CF a few years ago with tactical helicopter detachments being deployed to Bosnia. The question was asked whether the practice of having the senior pilot (at this point, often a young and inexperienced major) command the detachment should be replaced by having the senior major (usually an engineering officer) command the detachment; however, no change to the policy was implemented.[27]

While there is no detail about the results of this First World War leadership experiment, it would be interesting to pursue further. However, I would guess that it was a dismal failure because of the requirement for an effective squadron commander to demonstrate both technical skill and heroic leadership, as shown by the example that follows.

Heroic and Technical Leadership in the Air Force. The best squadron COs in both world wars were bold, skilled airmen who led by example. Those who were most admired carried out their orders intelligently and used their expertise to minimize the risks to the lives of their charges.[28] Sometimes exceptional technical skill was required to do this.

The unit of Victoria Cross winner Lanoe Hawker was the first to be equipped with DH2 aircraft, which had been rushed into service to counter the "Fokker scourge." The DH2 suffered from a number of manufacturing and technical problems, and it was soon dubbed the "Spinning Incinerator" by the pilots who flew it. On 13 February 1916, two of Hawker's best pilots were killed in accidents involving spins on their own side of the lines. Rumours quickly circulated among his pilots that these machines were death traps. A complete collapse in squadron morale seemed imminent, and Hawker had to act quickly. Immediately after the fatal accidents, he took a DH2 up on his own and recovered from every possible spin condition. He then explained the proper manoeuvres to his pilots, and they all practised them until they were proficient in spin recoveries. After that, while Hawker was in command, his squadron did not lose another flier from spinning into the ground. Thus, a potentially serious morale problem was avoided by a CO demonstrating his flying competence and by taking a personal risk.[29]

This next example of air force leadership is taken from the Second World War to show that while the principle was the same, different circumstances called for different actions. In terms of total losses, Bomber Command suffered grievously compared to other formations, on what has been called the "cutting edge of battle." Canadian rifle companies fighting the early campaigns in Italy, and British and American infantry in Normandy, experienced casualty rates of 50, 76, and 100 percent of unit strength respectively.[30] Bomber Command casualty rates for 1943 were 250 percent of unit strength.[31] During the Allied combined bomber offensive (1942–1945), 18,000 aircraft were lost; 81,000 British, Commonwealth and American fliers were killed; and combat casualties exceeded 50 percent of aircrew strength on average.[32] Naturally, aircrew leadership was a formidable challenge in these circumstances.

Unlike most of their First World War counterparts, RAF Bomber Command COs could not lead by being visually conspicuous to their followers. Most of their "ops" (operations) were conducted at night in loose bomber streams where crews might never see another aircraft. Therefore, Bomber Command leaders had to use novel methods to demonstrate heroic leadership and technical competence. The case of the RAF's 76 Squadron in 1943 is one such example.

Some COs got the derisive nickname "Francois" from their subordinates because they usually participated only in relatively safe raids on France. Not Leonard Cheshire. He deliberately elected to fly as second pilot "with the new and the nervous" on dangerous raids. In this way he demonstrated competence and risk-taking to his followers. By the end of the war, Cheshire had earned a Victoria Cross, three Distinguished Service Orders and a Distinguished Flying Cross and had become "a legend." His replacement had a much different experience. Rarely flying on dangerous ops and plagued with "bad luck" early returns, the new CO saw the unit's efficiency and morale deteriorate alarmingly. By the spring of 1943, 76 Squadron's early-return rate sometimes exceeded 25 percent of the aircraft dispatched. At the end of 1943 this CO was replaced. His successor, "Hank" Iveson, resumed the custom of the CO flying dangerous missions, and he broke up crews with persistent early-return records. This resulted in better unit performance, which significantly improved morale, but a CO had to be constantly alert to maintain it at a high level. When the squadron was re-equipped with the new Mark III Halifax, which had a "fearsome reputation for accidents," Iveson and his three flight commanders flew on the first operational mission with this aircraft to demonstrate their confidence in the squadron's equipment.[33]

The example of 76 Squadron shows how aircrew would follow charismatic leaders. Crews could not be driven to their tasks in Bomber Command; there were too many ways to shirk the tasks, especially on night operations, if crews felt their leaders were letting them down. For example, they could "deliberately sabotage" their aircraft to avoid going on ops;[34] they could "boomerang" (return early) or become "fringe merchants" (those who bombed on the edge of the target to avoid defences); and as the bombing campaign penetrated further into Germany in order to get above the defences, crews could jettison their bombs in the sea or over occupied Europe.[35] Good Bomber Command leaders inspired their men to press home the attack in the face of overwhelming odds against survival.[36]

An RAF wartime leadership study, which I have discussed in more detail elsewhere, highlighted the three main leadership lessons found in the examples given above: (1) no one type of personality ensured good leadership, but good leaders behaved in certain ways; (2) before a new squadron commander could be an effective leader, he had to first of all demonstrate his operational flying ability; and (3) leaders had to share the risks with their subordinates by going on "difficult raids," especially "when losses [were] heavy, or morale low." The importance of leadership, according to this study, was such that "the fortunes of the squadron" were often described in terms of its COs. One station commander remarked that cases of lack of confidence in leaders "usually occur in epidemics, and when an epidemic occurs, it is usually due to a bad squadron or flight commander." In one case, when "it became known that a squadron commander wouldn't fly operationally," five cases of lack of moral fibre (LMF) occurred in the first fortnight. Men cracked "because they had no confidence" in their leaders.[37]

Even at the higher levels of air force leadership, risk had to be shared from time to time for commanders to have credibility with the crews. On Bomber Command's first 1,000-plane raid (30–31 May 1942) casualties were expected to be high,[38] and one station commander is quoted as having said: "The C-in-C says you will spread apprehension and despair throughout Germany.... I have therefore delegated my duty in the Ops Room...in order to satisfy my pleasure in observing your firework display from the rear turret of 'A' Flight Commander's aircraft."[39] By choosing to fly on what was expected to be Bomber Command's most dangerous raid of the war to date in the most hazardous position of the aircraft, this station commander was an inspiration to his crews, and on this raid at least one Group Commander (two-star general equivalent) and several other station commanders flew with their men.

Based on the historical record, I suggest that there were several types of wartime air force leadership, each with a different balance between technical and heroic leadership styles. At the unit level, good flight commanders exhibited high levels of technical and heroic leadership. At the squadron commander level, the requirement for technical leadership started to diminish but the requirement for heroic leadership was still high. At the formation level (from colonel equivalent up to two-star generals, wing commanders up to group commanders), the requirement for technical leadership in the form of aircraft-handling skills diminished, but occasional heroic leadership was still necessary to inspire confidence in aircrews. At the highest level of air force command, technical leadership (in the sense of flying skills) was not important at all; physical risk-taking also was not required, but these leaders were expected to risk their careers for the welfare of their crews. For example, Trenchard and Sir Arthur "Bomber" Harris were not expected to fly at all — in fact, Harris almost never left his HQ, or visited units — but both were perceived to demonstrate exceptional concern for the welfare of their subordinates, especially in getting resources (such as new equipment and more personnel) for them. Despite the fact that Harris was nicknamed "Butch" (for Butcher) by his crews — not because of what he was doing to Germany but because of what he was doing to them — veterans of Bomber Command showed exceptional loyalty to Harris after the war. Most of them believed that he had done everything he could to ensure their welfare and that his strident advocacy of Bomber Command had caused him to be slighted in the post-war honours list.[40]

The leadership examples given above suggest that perhaps the greatest differences between army and air force leadership lie at the lower levels. In wartime, flight and squadron commanders were expected to demonstrate a type of heroic leadership that Keegan attributed to Alexander, but it was based on specialized knowledge and skills, particularly the ability to fly and fight an aircraft. As officers in the army and air force achieve senior rank, however, their masks of command may start to look increasingly similar. Another similarity between army and air force leadership is the assumption that it is more appropriate for certain occupations, such as the combat arms (or the aircrew, in the case of the air force), to provide the bulk of the leaders in the organization. In the air force, this could be called "the cult of the pilot."

The Cult of the Pilot. At the beginning of the First World War, a person's military occupation (such as pilot) did not automatically determine leadership status in the RFC/RAF, as we have seen. Furthermore, as the war became more technically complex, new occupations were created, such as armaments, photography and wireless, to complement the earlier technical trades of riggers and fitters and the support trades like administration, motor transport and stores. With the huge increase in size of the British air services, from just over 2,000 men in the RAF in

1914 to over 290,000 men and women in uniform in 1918,[41] all of these specialties developed their own officer and NCO corps that were responsible for overseeing the technical expertise necessary to keep the flying services operational.[42]

After the war the RAF and the Canadian air services were drastically reduced in size. In terms of leadership, this meant that most specialists were demobilized and that almost the entire officer corps consisted of pilots to ensure as many of them as possible were available to fly in the minuscule air forces of the inter-war years. One reason for this policy was that even constant peacetime flying took its toll due to stress, and ground jobs were generally reserved for pilots who were taking a break from flying. In addition to their flying duties, career air force pilots were expected to specialize in another trade, for example, armaments, photography, or navigation.[43] At the time the RCAF referred to pilots as "general list" officers (the RAF still refers to its aircrew as "General Duties" officers) because they were not viewed as specialists but as people who could fly and still do ground jobs, as opposed to specialists like "engineering officers" who could perform only ground duties. So the inter-war years saw the rise of the cult of the pilot where Trenchard and his Canadian proteges enforced his wartime dictum that pilots were more than airborne chauffeurs and would fill virtually all command positions.[44]

For the Canadian air force, this changed with the Second World War and the huge expansion of the RCAF. In 1938 its strength was 1,150 all ranks; at the end of 1943 its strength reached a wartime peak of 206,350, and of that number, 46,272 were overseas. In addition, the British Commonwealth Air Training Plan (BCATP) furnished 44 percent of the 340,000 Commonwealth aircrew trained between 1939 and 1945. Most of the training for the RCAF's expansion was done in Canada. With the outbreak of the war Canada went from training 10–20 pilots per year to training over 5,000 aircrew of all types *per month* under the BCATP.[45] Aircrew usually held a minimum rank of sergeant, and commissions in the RCAF were granted on the basis of marks in flying training, with about one half of the pilots being commissioned initially.[46] By the end of the war, virtually all Canadian aircrew were commissioned at the end of training. As in the inter-war years, pilots held most of the major command positions. But by 1942 the high loss rates and trouble finding enough good leaders among the pilots led to a fierce debate in the RAF and RCAF over whether other aircrew trades could command squadrons and flights. Necessity provided the answer, and soon navigators and a few wireless-operator air gunners and other aircrew trades were given command positions.[47]

As in the First World War, the massive expansion of the technical trades led to the reappearance of the officer and NCO hierarchies that had almost disappeared after that war. Even so, it was the aircrew who did most of the dying. While ground crew outnumbered aircrew by about five to one, 94 percent of the RCAF's fatal casualties were aircrew.[48] After the Second World War, despite the continued existence of most of the technical branches and some of their officers and senior NCOs, there was a return to the cult of the pilot (which has persisted in the Canadian air force until relatively recently) when officers from other occupations, for example, air navigators, could command squadrons, an aerospace engineering officer could become Assistant Chief of the Air Staff, and most recently, an air navigator could become Chief of the Air Staff.[49]

The dominance of pilots in the air force command structure has had a number of implications for air force leadership. While a great deal more research is needed in this area, it might be fair to characterize air force ground-crew leaders as requiring technical leadership skills more than heroic skills, as shown in Figure 1.2.

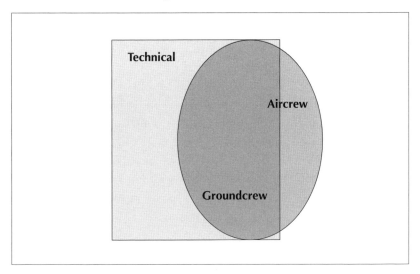

FIGURE 1.2. AIR FORCE LEADERSHIP

Besides the degree of technical leadership style used by aircrew and ground-crew officers, there are other significant differences between them. Unlike most ground-crew officers who, like junior army officers, often lead small sections as part of their first job, aircrew rarely get the chance to lead until they reach the rank of major and became flight commanders. Furthermore, most of their leadership experience is gained with peers and fellow officers and not with airmen or airwomen. This also means that aircrew do not receive mentoring from senior NCOs in their first command appointments in the same way that ground-crew officers do. Therefore, as they acquire their leadership skills, most senior air force leaders have very different formative experiences from army officers. One would expect that this would lead to very different approaches to leadership in joint-command situations — something that was noted by those who observed the genesis of pilots in leadership roles in joint operations in a maritime setting.

Naval Leadership[50]

The possibility of fliers exercising leadership roles in maritime operations seemed to be remote in 1918 because Britain's leading medical journal had "...ascertained that really good pilots are almost invariably bad sailors."[51] Some may still hold this view today, but, as in the air force, technology and those who direct its use have been integral to a navy's fighting ability. In Nelson's navy, for example, the man o'war was the most complex weapons system of its time, and it was supported by a highly sophisticated technical infrastructure.[52]

For this study of leadership there are three important points that come from naval history. First of all, technical competency was a crucial attribute for officers. Unlike the British army where commissions were purchased until 1871,[53] the Royal Navy (RN) always insisted that officers pass rigorous exams before being commissioned and before being qualified to assume command of a vessel.[54] This remains true today as naval officers are still required to submit rigorous command exams if they wish to become the master of a vessel.[55]

Second, in the eighteenth and early nineteenth century the RN only had two operational commissioned-officer ranks — lieutenant and captain — and advancement was largely by merit.[56] Promotion (or demotion) occurred by moving up (or down) to more (or less) demanding ships. For example, a captain might start his command career with a small sloop (18 guns) and advance all the way to command a first-rate ship of the line (100 guns), but he still retained the rank of captain. This reflected the realities of leadership in the naval world and is still reflected in the ranks of some European navies and in the French translations of CF naval ranks.[57] The RN rank system gave a great deal of flexibility in employing officers, and is perhaps more suited to today's highly technical and demanding leadership environment than is the multi-layered system in use today.

Third, all those who held the Queen's commission in the RN (that is, lieutenants and captains) were professional seamen and professional warriors. The division between professional seamen and others was evident in the crew's organization. The ship's company was divided into two groups: those who stood watch (the professional seamen) and the "idlers" (all those technicians who supported the ship and its company, such as armourers, cooks, the chaplain, the barber-surgeon). The idlers usually comprised less than 10 percent of the crew of about 800 on a first-rate ship of the line.[58]

The culture of the navy, including the Royal Canadian Navy (RCN), put a premium on both technical ability (seamanship) and career status (professional naval officer versus other officers) and formalized them in the rank insignia, which were different for each category of officer in the Canadian naval service in the Second World War, so that all, particularly the many wartime newcomers to the naval service, could be made aware of the cultural assumptions shown in Table 1.1 below.

Royal Canadian Navy	professional sailor	professional warrior
Royal Canadian Naval Reserve	professional sailor	amateur warrior
Royal Canadian Naval Voluntary Reserve	amateur sailor	amateur warrior

TABLE 1.1. CULTURAL ASSUMPTIONS IN THE CANADIAN NAVAL SERVICE IN THE SECOND WORLD WAR

When the Royal Canadian Naval Voluntary Reserve (RCNVR) was created in 1923, it adopted the "wavy navy" stripes of the Royal Naval Voluntary Reserve (RNVR) for its officers, and some of RNVR's traditions are perpetuated in the CF's naval reserve today. As a wartime expedient, the Royal Canadian Naval Reserve (RCNR) was created from the pool of merchant seaman in Canada, and its rank insignia was also distinctive. The rank insignia worn by each officer group in the Canadian naval service made clear the stereotypes attached to each class of naval officer as shown in Table 1.1. The RCNR was seen as merchant mariners who were professional sailors but who had no experience with war at sea. The RCNVR was perceived to be made up of "landsmen" who had neither experience at sea nor experience with war at sea. Some yachtsmen-dilettantes from Toronto's social elite fit this stereotype of poor sailors and bad officers, and in the press (including their yachting magazines) during the war they publicly criticized the Navy and the way it treated reservists. The reality, however, was that by 1944 many of these reservists had more operational time at sea and more time in contact with the enemy than most senior RCN officers had. From a cultural perspective, some of these same wartime naval issues are found in the Navy today, especially between Class A and B Reserves.[59]

The Tradition of Mutiny in the Canadian Navy. Little has been published on Canadian naval leadership, but some scholarly insights on the subject can be found in recent works on naval mutinies, or "incidents," as they were often referred to in the navy.[60] The tradition of mutiny in the Canadian navy comes from the RN (which probably adapted it from the British form of crowd social protest common in the eighteenth century).[61] Therefore, the stereotypical mutiny seen in films like *Mutiny on the Bounty* was rare in the navies of the British Empire and Commonwealth. The most common form of mutiny was the "industrial action" or sit-down strike to right specific wrongs. The form that mutinies usually took can be imagined from these unwritten rules of mutiny in the RN: "(1) No mutiny shall take place at sea or in the face of the enemy; (2) no personal violence may be employed (although a degree of tumult and shouting is permissible); and (3) mutinies shall be held in pursuit only of objectives sanctioned by the traditions of the Service." As long as they followed these rules, mutineers usually were not treated harshly. Most often, their grievances were recognized as legitimate by senior officers, and it was not unusual for the captain of a ship and/or his executive officer to be replaced, especially if their technical abilities (seamanship) were suspect.[62] Even Nelson himself expressed support for the actions of some of the sailors in the Great Mutinies of 1797. Writing to the Duke of

Clarence, who was the third son of King George III and who had served with Nelson in the navy, Nelson said: "I am not surprised that Your Royal Highness should have felt all the Agony of suspense during the late extraordinary Acts at Portsmouth… But to us who see the whole at once we must think that…it has been the most Manly thing I ever heard of, and does the British Sailor infinite honour."[63]

There were a number of reasons mutinies occurred with some regularity in the navies of the British Empire and Commonwealth. Primarily, the divisional system (started in the RN in the 1790s), the official way of dealing with grievances on board ship, did not work very well because it seemed threatening and inefficient to many sailors.[64] This perception existed because the divisional system was adequate for dealing with petty grievances, but it was not able to handle bigger problems like incompetent leaders. One reason for the weakness of the divisional system was a general lack of communication down the chain of command. For the sailors, mutiny was a risky but proven method of dealing with serious problems that they felt could not be resolved in any other way. Senior non-commissioned members of the crew, if they saw the grievances as legitimate, supported the mutinies by taking no action, tacitly encouraging them, or openly leading them (as happened in the Great Mutinies of the RN in 1797), depending on the circumstances.[65] These issues are relevant to the Canadian Navy today as some officers feel that the divisional system needs improvement to work efficiently and that the Canadian Navy still does not fully understand many of these issues related to discipline and leadership.[66]

Armies and air forces have also had their share of mutinies, but they are generally not as well documented as naval mutinies. For example, in January 1919 men of the newly formed Canadian Air Force refused to work, very much in the tradition of the naval mutiny, to protest what they perceived to be intolerable living conditions and poor leadership.[67] The most well-known recent case in the Canadian Army, that of Matt Stoppford, was not, as some have claimed, an atypical reaction by cowards who were trying to poison their leader, but really a type of mutiny that has been fairly well documented in other military forces. The actions of the men who put "noxious substances" into the unfortunate Matt Stoppford's coffee can best be understood as a form of protest against a leader they believed was unnecessarily risking their lives.[68]

Based on the discussion above, a model of naval leadership has been hypothesized as shown in Figure 1.3. The model has similarities with the air force model shown above, but there are differences due to the navy's culture, for example, the overlap between the technical sea trades and the idlers. The greatest similarity is the large role that the technical style of leadership plays in naval leadership. In addition, this model illustrates the importance of the technical style of leadership for naval officers. Again, more research is required in this area, but it appears that for naval officers, like aircrew, technical competence is at least as important as leadership competence because without technical competence these officers could not normally hold a leadership position for long.

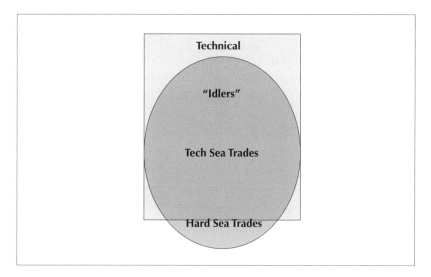

FIGURE 1.3. NAVAL LEADERSHIP

History and Leadership Models

The penultimate section of this chapter proposes some ways of looking at service leadership in this country. Since very little research has been done on the topic, much of it is speculative and based on personal observations, but it may stimulate some interest in these topics for future research. The section starts with an examination of differing hierarchies of loyalty in the three services, because the order in which military personnel perceive their loyalties to lie may shed some light on differences in service leadership.

It appears that because people change units (ships and squadrons) frequently in the navy and air force, their hierarchy of loyalty is (1) service (navy or air force), (2) job/occupation (for example, maritime engineer, pilot), and then (3) unit (ship or squadron). There is some culturally based evidence for this assumption: in the pre-unification RCN and RCAF, the cap badges were the same for all officers in each service.

For the army, I offer an alternate interpretation to what is found in much of the Canadian literature on army leadership, where it is assumed that the regimental system is at the heart of the army culture with a hierarchy of loyalty going from (1) regiment, (2) branch (such as infantry, artillery) and then to (3) the army as a service. I suggest that the traditional interpretation is only really true for the infantry in the Canadian context. The armoured corps is more problematic because its members routinely re-badged to serve in Canadian Forces Europe up until 10 years ago, although today they tend to remain more within their regimental families. Other groups, I suggest, owe their first loyalty to their job/branch/occupation (for example, gunner, engineer, signals) because they do not have regiments in the same sense as the infantry has. This produces a loyalty hierarchy as follows: (1) job/branch, (2) service and (3) unit. Such hierarchy of loyalty bears some resemblance to the navy and air force hierarchies, I suggest, because of the relatively high technical leadership component found in the cultures of these "other" army subcultures. This gives rise to a model, shown in Figure 1.4, where the technical style of leadership is a greater component of the army leadership domain than is often acknowledged.

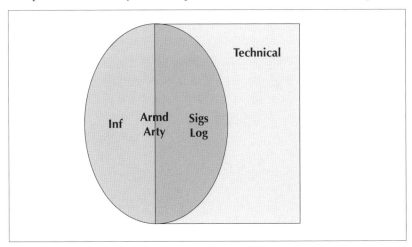

FIGURE 1.4. ARMY LEADERSHIP

Again, a great deal of research needs to be done before any of these tentative conclusions can be accepted as valid, but they may be a starting point for debate.

Next, I would like to use a familiar leadership model to summarize some current perceptions about Environmental leadership in the CF and to illustrate some of the strengths and weaknesses implicit in using models to describe military leadership.

Using a Leadership Model. I have chosen the Hersey-Blanchard model because it is taught to many officers in the CF, especially those who attend the Royal Military College of Canada, and it is seen by many to have direct application to military leadership.[69] The point to be made in Figure 1.5 below is that the technical leadership style tends towards S3 and S4 because followers are expected to be competent in their jobs and do not need to be told how to do them; rather, the leader's job is to encourage and facilitate high job performance.

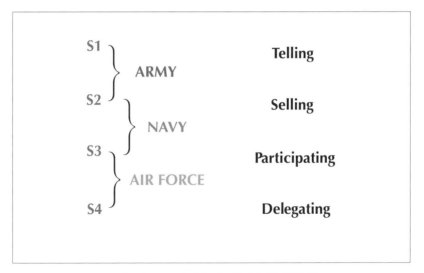

FIGURE 1.5. PERCEPTION OF PREFERRED LEADERSHIP STYLES

I have used the Hersey-Blanchard model for several purposes: first of all, to try to simplify the relationship among service leadership styles and to put in a visual format some commonly held perceptions about leadership in the CF; secondly, to show how a leadership theory might be applied to future leadership challenges; and finally, and I think most importantly, to demonstrate how models and theories must be used with extreme caution when applying them to reality. The drawbacks of using a theory or model to simplify very complex situations are evident in Figure 1.5, as many army leaders have used S3 and S4 leadership styles and this diagram grossly oversimplifies what we know happens in real life.

The lesson here is to "beware the tyranny of models." Many find it comforting to reduce reality to models, and it is useful to do this when appropriate. But I believe one job of historians involved in the study of leadership is to challenge the perceptions that models sometimes engender and to encourage their creators and those who would use them to remain wary of putting too much faith in them. History shows that models will fall in and out of favour based on how shifts in national and service cultures affect beliefs about their explanatory power. For example, the current fascination with transformational leadership as a leadership style, to the exclusion of other leadership styles in some quarters of DND, should serve as a caution. Alexander the Great could be seen as the epitome of the transformational leadership style, but he also used transactional leadership. Or one might characterize his leadership in altogether different terms, as Keegan did.

As we have seen, some of the complexities in the historical record of leadership defy neat categorization. However, in real life this complexity must be balanced with the need to bring some order to the data. To do this, armed forces will always need the advice of experts from various disciplines, but the sheer quantity of the data means that military professionals are often being inundated with expert advice, much of it conflicting. What are they to do in these circumstances?

I suggest that military professionals should evaluate competing views based on the following guidelines. First, demand clear explanations free of jargon; if clear explanations cannot be provided, be suspicious. Second, demand to see facts or data relevant to the issue at hand; be cautious about accepting data based on other circumstances or submissions that express strongly held beliefs with limited or no data. For example, we know that leadership has an important cultural component, but evidence gathered by researchers from one cultural domain, for example, the US Army, does not necessarily apply to another, for example, the Canadian Army.[70] Third, be cautious about accepting advice from disciplines that do not admit the shortcomings of their own methodology nor acknowledge the contributions of others. Be particularly wary of those who claim to have the "silver bullet" that will solve all your problems. Finally, always ask the question "Where's the evidence?" If relevant and comprehensible evidence cannot be provided by an advisor, then be wary of that advisor's advice.

In summary, when dealing with complex issues like leadership, military professionals need to consult many different kinds of experts and use a multi-disciplinary approach to balance the strengths and weaknesses of different disciplines. In a way, it is like building a house: many different skills are needed to build a satisfactory structure.

Conclusions

Understanding differences in leadership among the army, navy and air force has become increasingly important in an era where joint and combined operations predominate and integrated operations are becoming the norm. These differences, recognized in recent Canadian doctrinal publications on leadership and the profession of arms, are caused by cultural variations not only among nations but among services in any given nation. While there is a rather extensive literature on army leadership and much less of a literature on navy and air force leadership, there has been very little written about the differences in leadership among the services or how these differences might affect joint or integrated operations. This is especially true in Canada, where the military leadership literature is sparse.

In this chapter I contend that every service has different leadership expectations based on that service's mask of command. Even in the unified CF, where a significant amount of training and education is conducted in a joint environment, leaders spend their most formative years in a single service culture that shapes their views about what is an appropriate leadership style.

To examine service leadership differences I have used two generic types of leadership: heroic and technical. Heroic leadership, defined here as conspicuous sharing of risk with subordinates, appears to be common to combat leadership in all services. In addition to heroic leadership, I suggest that there is a second generic leadership style, which I have called "technical," that is found as a subculture, in different proportions, of each Environment's (or service's) culture.

Technical leadership is defined in this chapter as the ability to influence others to achieve a goal, based on the specialized knowledge or skill of the leader. Such leadership is particularly important to air force and naval officers, who must display technical competence before they will be accepted as legitimate leaders by their subordinates. Technical leadership also affects the way members of the army, navy and air force form attachments that are the foundation of cohesion and morale.

In each service there is a hierarchy of loyalties that influences how leadership can be exercised in each combat environment. In the Canadian Navy and Air Force the loyalty hierarchy appears to be (1) service (navy or air force), (2) job/occupation (such as maritime engineer, pilot), and then (3) unit (ship or squadron). For the Canadian infantry, and to some degree the armoured corps, it seems to be (1) regiment, (2) branch (infantry or armoured), and then (3) the army as a service. For other army branches, because of their relatively high technical leadership component, it may be (1) job/branch, (2) service, and (3) unit.

Much in these models is speculative in nature and serves more as an invitation to others to conduct research into these areas of inquiry than as a statement of any definitive findings. I see the most profitable future research as a co-operative, multi-disciplinary endeavour where researchers can bring the strengths of their disciplines to bear on these questions.

Recent Canadian doctrinal publications on leadership and the profession of arms have provided a good base for understanding leadership in the CF. However, until we know a great deal more about differences in Environmental (or service) culture and leadership, Canadians should be cautious about using longstanding stereotypes of service leadership and accepting conclusions based on foreign data. In the interim the Canadian military can rely on its proven strengths of joint education and training that expose officers of a given service to the other service cultures, at least in a superficial way. But until we can understand the Canadian service cultural mosaic and its impacts on leadership in the CF much more clearly than we do now, we should proceed with great prudence in adopting any enduring leadership doctrine based on prevailing, unverified cultural myths and assumptions.

NOTES

1   This essay is based on "The Masks of Command: Leadership Differences in the Canadian Army, Navy, and Air Force," a paper given at the Conference on Leadership in the Armies of Tomorrow and the Future, Kingston, Ontario, 6–7 February 2002, and also at the Inter-University Seminar on Armed Forces and Society, Kingston, Ontario, 25–27 October 2002.

2   *Combined* in this context means activities, operations, organizations et cetera between two or more forces or agencies of two or more allies. Canada, Department of National Defence (DND), Glossary, in *Canadian Forces Operations*, B-GG-005-004/AF-000 (December 2000).

3   *Joint* is defined as activities, operations, organizations et cetera in which elements of more than one service of the same nation participate. Glossary, in *Canadian Forces Operations*.

4   Before unification Canada had three separate services: the Royal Canadian Navy (RCN), the Royal Canadian Air Force (RCAF) and the Canadian Army. After 1 February 1968, when the Canadian Forces Reorganization Act took effect, all Canada's armed forces were unified into a single service, the Canadian Armed Forces (CF). While the RCN, the RCAF and the Canadian Army no longer existed as legal entities, people often referred to the navy, air force and army in everyday usage. However, to emphasize the point that Canada no longer had three services, DND bureaucrats coined the rather awkward term *environment* (based on the environments in which the sea, air and land components of the CF operate) to describe these components. Since there is only one military service in Canada today, the CF, official DND publications sometimes use the noun *environment* and the adjective *environmental* when referring to the sea, air and land components of the CF. Nonetheless, the terms *Canadian Navy, Canadian Air Force* and *Canadian Army* are creeping back into official usage.

5   DND, *Duty with Honour* (Kingston, ON: Canadian Defence Academy, 2003), 51.

6   *Duty with Honour*, 59.

7   *Duty with Honour*, 74.

8   *Duty with Honour*, 25.

9   *Duty with Honour*, 74.

10   DND, *Leadership in the Canadian Forces: Conceptual Foundations* (Kingston, ON: Canadian Defence Academy, 2005), 4.

11   *Duty with Honour*, 73.

12   B.H. Liddell Hart, *Why Don't We Learn from History?* (New York: Hawthorn, 1971), 15.

13   John Keegan, *The Mask of Command* (London: Jonathan Cape, 1987), 113, 118–19.

14   Keegan, *The Mask of Command*, 10.

15   Carl H. Builder, *The Icarus Syndrome: The Role of Air Power Theory in the Evolution and Fate of the US Air Force* (London: Transaction Publishers, 1994), 5–7.

16   Walter F. Ulmer, Jr., et al., *American Military Culture in the Twenty-First Century* (Washington, DC: CSIS Press, 2000), 13.

17   See Brian D. Wheeler et al., "Aerospace Doctrine?" in *Air Power at the Turn of the Millennium*, eds. David Rudd et al. (Toronto: Canadian Institute of Strategic Studies, 1999), 141–77, for an overview of the problems with Canadian aerospace doctrine. A more detailed account may be found in John Westrop, "Aerospace Doctrine Study," unpublished report dated 30 April 2002, copy at the Canadian Forces College library.

18   Canadian military culture is discussed in detail in Allan English, *Understanding Military Culture: A*

Canadian Perspective (Montreal and Kingston: McGill-Queen's University Press, 2004).

19   Some of the concepts discussed here have been developed further in an essay titled "Leadership and Lack of Moral Fibre in Bomber Command, 1939–1945: Lessons for Today and Tomorrow," in *Historical Perspectives of Mutiny and Disobedience, 1939 to Present* (Kingston, ON: Canadian Defence Academy Press, in press).

20   S.F. Wise, *The Official History of the Royal Canadian Air Force,* Vol. 1, *Canadian Airmen and the First World War* (Toronto: University of Toronto Press, 1980), 597.

21   Wise, *Canadian Airmen and the First World War,* 597; and A. C. Critchley, *Critch!: The Memoirs of Brigadier-General A.C. Critchley* (London: Hutchinson, 1961), 88.

22   Andrew Boyle, *Trenchard* (London: Collins, 1962), 96–141; citation from page 199.

23   Ibid., 33.

24   Wise, *Canadian Airmen and the First World War,* 100–1, 118.

25   Allan D. English, *The Cream of the Crop: Canadian Aircrew, 1939–1945* (Montreal and Kingston: McGill-Queen's University Press, 1996), 65.

26   Walter Raleigh, *War in the Air,* vol. 1 (Oxford: Clarendon Press, 1922), 438.

27   See Allan English, "Leadership and Command in the Air Force: Can Non-Aircrew Command Flying Squadrons?" in *Proceedings: 6th Annual Air Force Historical Conference,* ed. Office of Air Force Heritage and History (Winnipeg, MB: Air Force History and Heritage, 2000), 79–86, for a discussion of this issue.

28   See for example Jean Beraud Villars, *Notes of a Lost Pilot,* trans. Stanley J. Pincetl, Jr., and Ernest Marchand (Hamden, CT: Archon Books, 1975), 97, 146–47, 169, 200; Roger Vee (Vivian Voss), *Flying Minnows: Memoirs of a World War I Fighter Pilot from Training in Canada to the Front Line, 1917–1918* (London: Arms and Armour Press, 1976), 239; Curtis Kinney with Dale M. Titler, *I Flew a Camel* (Philadelphia: Dorrance, 1972), 88; and English, *Cream of the Crop,* 65.

29   Tyrrel Mann Hawker, *Hawker, V.C.* (London: Mitre Press, 1965), 125, 129, 135, 140–43. This example was first presented by Allan English, "Leadership and Lack of Moral Fibre in Bomber Command, 1939–1945," in *The Evolution of Air Power in Canada,* vol. 1, eds. William March and Robert Thompson (Winnipeg, MB: Air Command History and Heritage, 1997), 67–75.

30   John A. English, *On Infantry* (New York: Praeger, 1984), 138.

31   Allan English, *Cream of the Crop,* 101.

32   Mark K. Wells, *Courage and Air Warfare* (London: Frank Cass, 1995), 2.

33   Max Hastings, *Bomber Command* (New York: Dial Press, 1979) 247–48, 252. This example was first presented by Allan English, "Leadership and Lack of Moral Fibre in Bomber Command, 1939–1945."

34   Some examples are given by Hastings, *Bomber Command,* 248 (deliberately fouling the magnetos while running up the engine); and Norman Longmate, *The Bombers* (London: Hutchinson, 1983), 184 (tampering with gun-turret hydraulic systems).

35   The number of bombs "jettisoned" during the Battle of Berlin has been described as "enormous." Charles Webster and Noble Frankland, *The Strategic Air Offensive Against Germany, 1939–1945,* vol. 2 (London: HMSO, 1961), 195–56. Bomber Harris was aware of these problems. The policy of having tour lengths defined by successful sorties, where possible confirmed by photos taken at bomb release, was designed to discourage "fringe merchants" and "boomerangs." Charles Messenger, *"Bomber" Harris* (London: Arms and Armour Press, 1984), 90; and John Terraine, *The Right of the Line* (London: Hodder & Stoughton, 1985), 524.

36   Hastings, *Bomber Command,* 247–48, 252.

37   Allan English, *Cream of the Crop,* 93–7. Quotes from C.P. Symonds and Denis Williams, "Personal Investigation of Psychological Disorders in Flying Personnel of Bomber Command," FPRC Report 412 (f), August 1942, in Air Ministry, *Psychological Disorders in Flying Personnel of the RAF,* Air Publication 3139 (London: HMSO, 1947), 53–4.

38   Royal Air Force casualties were at a record high: 43 aircraft lost (4 percent of the 1,047 attacking). Martin Middlebrook and Chris Everitt, *The Bomber Command War Diaries* (London: Penguin, 1990), 272.

39   Longmate, *The Bombers*, 221.

40   See Dudley Saward, *"Bomber" Harris* (London: Cassell, 1984), 324–34, for a spirited defence of Harris's reputation.

41   H.A Jones, *War in the Air: Appendices* (Oxford: Clarendon Press, 1937), Appendix XXV.

42   Denis Winter, *The First of the Few* (Athens, GA: University of Georgia Press, 1983), 110–20, gives a description of some of the work performed by ground crew in the First World War. H.A. Jones, *War in the Air*, vol. 5 (Oxford: Clarendon, 1935), chap. 8, gives an account of how training for the new trades was conducted.

43   W.A.B. Douglas, *The Official History of the Royal Canadian Air Force*, vol. 2, *The Creation of a National Air Force* (Toronto: University of Toronto Press, 1986), 145.

44   This point was brought to the attention of American officers during a May 1917 visit to RFC Canada, where they were told that the pilot was not a "flying chauffeur" but a "modern cavalry officer" or a "knight of old." Hiram Bingham, *An Explorer in the Air Service* (New Haven, CT: Yale University Press, 1920), 16–17.

45   F.J. Hatch, *Aerodrome of Democracy* (Ottawa: Directorate of History, DND, 1983), 101, 194, 205; and Douglas, *The Creation of a National Air Force*, 192, 226–7, 247.

46   Douglas, *The Creation of a National Air Force*, 221; and Allan English, *The Cream of the Crop*, 120–1.

47   See PRO AIR 14/290, particularly BC/C.23068 and attached minute sheets for details. Harris's approval for this new policy is at minute 43.

48   C.P. Stacey, *Arms, Men and Governments* (Ottawa: Queen's Printer, 1970), 66, 305; and W.R. Feasby, ed., *The Official History of the Canadian Medical Services, 1939–1945* (Ottawa: Queen's Printer), 512.

49   The cult of the pilot was less predominant in maritime patrol and maritime helicopter squadrons where naval traditions had some influence and there was less concern with the occupation of squadron and flight commanders as long as they were aircrew. See James F. Johnson, "Air Navigators and Squadron Command Opportunities," *Canadian Forces Polaris* 2, no. 1 (1973), 40–1. See also Allan English, "Can Non-Aircrew Command Flying Squadrons?"

50   The subject of naval leadership in a Canadian context is discussed in detail in Allan English, Richard Gimblett, Lynn Mason and Mervyn Berridge Sills, "Command Styles in the Canadian Navy," Defence Research and Development – Toronto, Contract Report CR 2005-096 (31 January 2005), which is summarized in Richard Gimblett's chapter in this book.

51   "The Doctor and the Airman," *The Lancet*, 16 March 1918, 411.

52   The complexities of this weapon system are described in Roger Morriss, *The Royal Dockyards During the Revolutionary and Napoleonic Wars* (Leicester: Leicester University Press, 1983).

53   Correlli Barnett, *Britain and Her Army, 1509–1970* (London: Allen Lane Penguin Press, 1970), 307–9.

54   An excellent description of the naval officer corps at this time can be found in Michael Lewis, *A Social History of the Navy, 1793–1815* (London: Allen & Unwin, 1960). See pp. 141 and 267–8 on the competency issue.

55   See Allan English et al., "Command Styles in the Canadian Navy," 57–62, for a discussion of this issue in a Canadian context.

56   Midshipmen were officers under training, and admirals (flag officers) were promoted strictly by seniority.

57   For example, lieutenant-commander is translated as *capitaine de corvette*; commander is translated as *capitaine de frégate*; and captain is translated as *capitaine de vaisseau*.

58   See Lewis, *A Social History of the Navy*, 85–6, 270–80, for descriptions of the crew and their duties.

59   Richard Oliver Mayne, "'Equal Privileges for Greater Sacrifices': Insurrection in the Canadian Naval Reserve, 1942–44," unpublished paper, 25 February 2002. In the Canadian Navy in 2002, Class A reservists served about fourteen days per year in uniform and have been compared to the RCNVR. Class B reservists usually serve for a full year, often aboard minor coastal defence vessels, and have been com-

pared to the RCNR. I am grateful to Richard Mayne, one of my graduate students, for his insights into the Canadian Naval Reserve.

60   A literature on Canadian naval mutinies and, indirectly, leadership has begun to develop over the past few years. Some of the leading works are Richard H. Gimblett, "'Too Many Chiefs and Not Enough Seamen': The Lower Deck Complement of a Postwar Canadian Navy Destroyer — The Case of HMCS *Crescent*, March 1949," *The Northern Mariner* 9, no. 3 (July 1999), 1–22; Bill Rawling, "Only 'A Foolish Escapade by Young Ratings'? Case Studies of Mutiny in the Wartime Royal Canadian Navy," *The Northern Mariner* 10, no. 2 (April 2000), 59–69; Richard H. Gimblett, "What the Mainguy Report Never Told Us: The Tradition of Mutiny in the Royal Canadian Navy before 1949," *Canadian Military Journal* 1, no. 2 (Summer 2000), 87–94; Michael J. Whitby, "Matelots, Martinets, and Mutineers: The Mutiny in HMCS *Iroquois*, 19 July 1943," *Journal of Military History* 65, no. 1 (January 2001), 77–103; and Richard O. Mayne, "Protestors or Traitors? Investigating Cases of Crew Sabotage in the Royal Canadian Navy, 1942–1945," *Canadian Military Journal* 6, no. 1 (Spring 2005), 51–8.

61   See E.P. Thompson, *Customs in Common* (London: Merlin Press, 1991), particularly chap. 4, "The Moral Economy of the English Crowd in the Eighteenth Century," and chap. 5, "The Moral Economy Reviewed," 185–351, for a detailed explanation of this phenomenon. Thompson called this type of protest "the Moral Economy" because it kept the capitalist economy in balance.

62   See N.A.M. Rodger, *The Wooden World* (London: Collins, 1986), 237–44, for a description of mutiny in the RN during the age of sail. See also Gerald Jordan, "Admiral Nelson as a Popular Hero," in *New Aspects of Naval History*, ed. Department of History, U.S. Naval Academy (Baltimore: Nautical & Aviation Publishing, 1985), 117, for the actions of mutineers in the naval mutinies of 1797. See Rawling, "Only 'A Foolish Escapade by Young Ratings'?" 59, 69; Gimblett, "What the Mainguy Report Never Told Us," 93; and Whitby, "Matelots, Martinets, and Mutineers," 88, 99–103, for a description of how the Canadian naval service inherited the RN tradition of mutiny.

63   Nelson to the Duke of Clarence, unpublished letter, quoted in Dalya Alberge and Joanna Bale, "A Fool for Love: Hidden Passions of Lord Nelson," *Times Online* (4 April 2002), http://www.timesonline.co.uk.

64   See Christopher Lloyd, *The British Seaman, 1200–1860* (Rutherford, NJ: Farleigh Dickinson University Press, 1968), 234, and Rodger, *The Wooden World,* 216, on the origins of the divisional system.

65   See J.G. Bullocke, *Sailors' Rebellion* (London: Eyre & Spottiswoode, 1938), 211, and G.E. Manwaring and Bonamy Dobrée, *The Floating Republic* (London: Geoffrey Bles, 1935), 34–8, 262–3, for the role of the senior non-commissioned members of the crew in leading the Spithead part of the naval mutinies of 1797.

66   Gimblett, "What the Mainguy Report Never Told Us," 94.

67   Wise, *Canadian Airmen and the First World War*, 611.

68   The "Detailed Report of the Special Review Group Operation Harmony (Rotation Two)" dated 26 June 2000, http://www.forces.gc.ca/site/reports/harmony_2/index-e.asp, explains these issues in detail.

69   For a detailed description of this model and its applications see Paul Hersey and Kenneth H. Blanchard, *Management of Organizational Behavior* (Englewood Cliffs, NJ: Prentice-Hall, 1977).

70   Issues of leadership and culture in armed forces are discussed in more detail in English, *Understanding Military Culture*.

CHAPTER 2

CANADIAN NAVAL COMMAND STYLES[1]

Richard H. Gimblett

Recent research has indicated that the key to creating an adaptable and effective command and control (C^2) system is the establishment and nurturing of an organizational culture to support it. Innovation in large organizations is usually constrained more by the organization's culture than by technology. Western armed forces have not been particularly successful in this regard, and in some cases, dysfunctional military cultures appear to be frustrating the best intentions of commanders; witness the current debate over the post–Cold War revolution in military affairs that has accelerated the tendency to see C^2 in terms of technical systems, based on such concepts as *network-centric warfare*, sometimes referred to as *network-enabled operations*.[2] Therefore, a critical component of designing and implementing new C^2 systems is gauging current organizational culture, deciding on and articulating any necessary changes and then having the capability to implement them.

Working from the premise that the human dimension of command is critical in devising effective C^2 systems, this chapter examines naval command styles in the context of naval culture and organization from historical and contemporary perspectives. It analyzes the historical findings, using recent theories of command and control to gain further insight into naval command styles in general and Canadian naval command styles in particular.

A number of basic questions frame any investigation into the subject: Is there something that can be called a distinctly "naval" command style? If so, what distinguishes it from those styles that might be identified as typical of the other services, the army and the air force? Certainly, the phrase invokes images of a solitary officer pacing "his" private corner of the poop deck or, more modernly, the bridge. Are these images merely caricatures, or like all good stereotypes, are they grounded in a large measure of truth? Beyond those "big" questions are many smaller ones: Is any naval style (or styles) merely an extension of those of the other services? Or can a set of principles be identified that might be universally applied to all naval services regardless of national characteristics or government structure? These lead, ultimately, to the defining question: Is there a particularly "Canadian" style (or styles) of naval command?

Complicating any investigation of this sort is the astonishingly sparse critical literature regarding naval command. Titles such as *Command at Sea* tend to be little more than biographies of the "great admirals," usually an entirely subjective list of the author's favourites.[3] The literature gets even sparser when narrowed to a Canadian perspective, where a survey to identify works by or about Canadian naval commanders produces an embarrassingly short list of credible titles.[4] In essence, while neither this chapter nor a larger study can claim to be definitive, any attempt, by definition, is both original and a contribution to the field of study.

The first major contribution arises from the need for a unifying framework to lend coherence to the study. The wide-ranging input from diverse disciplines, as well as from practitioners of naval command, results in a somewhat eclectic view of naval command. In some respects, navies have clung to tradition more so than have armies or air forces, and this propensity, combined with the decades-long lives of ships, has made their organizational cultures more resistant to change than the other services. On the other hand, navies (including the Canadian Navy), by the nature of the environment in which they operate, usually have been at the forefront of technological change.

As the study progressed, it became apparent that one means of understanding differences in military command styles is to observe the influence on the way command is exercised in the different contexts of the three principle factors: environment (sea, land or air), technology (a major control mechanism for exercising command) and culture (service, organizational and national). This "environment-technology-culture" triad can explain why commanders in different services (army, navy and air force)[5] may react differently to the same circumstances, and why commanders from different nations may also react differently to the same circumstances.

Leadership, Management and Command in the Canadian Forces

Understanding differences in leadership and command among the army, navy and air force has become increasingly important in an era where joint and combined operations predominate. These differences are caused by differences in national and service cultures that vary not only among nations but also among services in any given nation. Even in the unified CF, where a significant amount of training and education is conducted in a joint context, leaders spend their most formative years in a single service culture that shapes their views about what is an appropriate leadership style. For example, unlike army and air force leaders, naval leaders must live and work in close confines with their subordinates, and especially on long sea voyages, they find themselves in a leadership position without a break for months on end.

To begin, the concepts of leadership, command and management are often conflated in the literature and in practice, and the position of ship's captain (or commander) epitomizes the problem of distinguishing between them. The latest CF doctrine on leadership and command underscores the issue:

> The inter-relationships and interconnectedness of command, management, and leadership *functions* often make it difficult to disentangle the command, management, and leadership *effects* achieved by individuals in positions of authority. Hence favourable results tend to be attributed to extraordinary leadership even when they may, in fact, be the result of command or management skills, some combination of all three, or other factors — including luck.[6]

For the purposes of this chapter, the broad definitions employed in *Leadership in the CF: Conceptual Foundations* pertain. *Leadership* is viewed as an influence activity potentially done by anyone, *command* is viewed as a creative and purposeful act reserved for those with legitimate authority, while *management* (especially in the sense of resources) is a subordinate but necessary complement to leadership and command.[7]

The complexity of the relationships among command, leadership and management is further complicated by the fact that they can be exercised at various levels of conflict. For the sake of simplicity, command in the Canadian Navy, as with practically all other navies, unfolds at three essential levels: the strategic level (national headquarters, sometimes referred to as *admiralty* after the British practice[8]), the operational level (ashore headquarters and higher formation level at sea, such as a task force with theatre responsibilities) and the tactical level (the individual ship unit or small groupings, generally now referred to as the *task group*).

Since command and leadership are inextricably intertwined, it is very difficult to separate behaviours into neat categories associated with each, and a separate discussion could be devoted to the purpose. Therefore, the reader is directed to *Leadership in the CF* as providing the authoritative CF theoretical foundation for such concepts as sources of a leader's power, and leader characteristics and influence behaviours; leadership at the tactical and operational levels of command; and command at the strategic level. As for command in a military context, theoretical study is still immature. One of the most rigorous developing frameworks that is culturally compatible with Canadian concepts of command is one put forward by Ross Pigeau and Carol McCann of Defence Research and Development Canada, Toronto.[9] Returning to base principles, they first distinguish the concept of command from the concept of control and then re-link the two concepts in a new understanding of C^2, giving new definitions to key terms as follows: *command* is the creative expression of human will necessary to accomplish the mission; *control* is those structures and processes devised by command to enable it and to manage risk; and C^2 is the establishment of common intent to achieve co-ordinated action. Within these concepts, they propose that command capability can be described in terms of three independent dimensions (*competency, authority* and *responsibility*), and they distinguish commander's intent as the product of two components: *explicit intent* (that part which has been made publicly available through orders, briefings, questions and backbriefs) and *implicit intent* (that part derived from personal expectations, experience due to military training, tradition and ethos, and from deep cultural values). As will be seen from the discussion below, all of these concepts resonate deeply in any attempt to comprehend naval command styles.

Arguably, it is at the lower tactical level that the distinction between leadership and command is most blurred, where the naval commander has the most frequent occasion for direct "influence" interaction with subordinates. However, the limited research into the subject again is a bar to deeper understanding. One promising avenue of American research looks to the cultural differences among the services. As noted in the previous chapter, Carl Builder contends that the touchstone of the US Army's organizational culture is the art of war and the profession of arms; in other words, concepts and doctrine are the glue that unifies the army's separate branches. For the US Navy, the heart of its organizational culture is the navy as an institution based on tradition, plus a maritime strategy, which provide coherence and direction to the navy. The US Air Force, in contrast, he declares, has identified with platforms and air weapons rooted in a commitment to technical superiority, and it has transformed aircraft or systems into ends in themselves, with serious repercussions. Writing in the early 1990s, Builder maintained that because the US Air Force had no integrating vision like the US Army's AirLand Battle or the US Navy's Maritime Strategy, it had conceded the intellectual high ground to the other services, particularly the Army.[10]

Similarities to Canadian service culture can be seen in Builder's model. For example, the Canadian Army invests a great deal in doctrine; the Canadian Air Force invests very little and remains focussed on platforms; and the Canadian Navy, with its deep-rooted traditions and new maritime strategy *Leadmark*,[11] exhibits many cultural similarities to its American analogue. However, beyond these basic similarities with the American services, Canadian military culture is based on its own historical experience.

A useful model of naval leadership emerges from recognition of the large role that technology, as integral to a navy's fighting ability, has played in naval culture. The land-centric context of army operations has given rise to the traditionally accepted "heroic" style of leadership (notable for its conspicuous sharing of risk with subordinates) as the basis for most models.[12] However, there is an emerging understanding that in technically oriented services such as the navy and the air force, technical competence is at least as important as leadership competence because without technical competence these officers could not normally hold a leadership position for long. *Technical leadership* is defined as the ability, based on the specialized knowledge or skill of the leader, to influence others to achieve a goal. It is exercised either by leaders (for example, pilots) who must be able to actually do the same job as their subordinates, or by leaders who must have a significant specialized knowledge of the jobs that their subordinates perform (for example, the seamanship skills of the naval officer). This type of leadership is critical in the navy and air force, where every second they are at sea or in the air, those on board ships and aircraft depend on technology (and by extension, the technical ability of the crews and their leaders), and not just their ability to fight, for their very survival, as noted by English in his chapter herein.

The next section will explore more fully the long historical development of this emphasis upon technical competence as the basis for naval command in a context especially pertinent to the Canadian Navy. It is instructive, however, to underscore certain enduring implications at this point. First, technical competence was a crucial attribute for naval officers, as seen from the insistence that they pass rigorous exams before being qualified to assume command of a vessel.[13] Second, advancement by rank was largely by merit, and promotion (or demotion) occurred by moving up (or down) to more (or less) demanding ships, a fact reflected most demonstrably in the French translations of CF naval ranks, equating to progressively larger vessels.[14] Third, the division between professional seamen and others was evident in the crew's organization into two groups: those who stood watch (the professional seamen, or "watchkeepers") and the "idlers" (all those technicians who supported the ship and its company, such as the purser, armourers, cooks, the chaplain, the barber-surgeon).[15] Clearly, naval culture puts a premium on both technical ability (seamanship) and career status (professional naval officer versus other officers). The navy is unique among the services in that it regularly subjects its commanders at the tactical and low operational levels to rigorous outside assessment of both their technical and leadership skills.

A final observation is required upon the environment within which naval leaders exercise their command. At the ship level, the naval leader is isolated in command and does not have to motivate the crew to follow in the same way that the army leader must. At the same time, while a ship is filled with specialists, each of whom offers information that is invaluable to the decision-making process, it would be a mistake to construe a ship's captain as head of an organization that operates on the basis of consensus building. At higher levels, the commander becomes progressively more remote, but the volume of information to manage increases exponentially with the expanse of the operational area to control (typically measured in the order of thousands of square kilometres) and concurrently across the full range of warfare domains (air, surface, sub-surface and, more recently, "information").

Although it is a truism that all navies share many things in common, ranging from the environment upon which they operate to the weapons with which they are equipped, those factors also can be the source of differences: for example, the tropical archipelagic waters upon which the Indonesian Navy operates demand a different type of vessel than the open-ocean, sub-arctic areas off our coasts demand; and the riverine vessels of the Ecuadorian Navy carry much smaller-calibre weaponry than our mostly larger vessels carry, and they have no need to counter air or submarine threats. Thus, such factors as the size of vessel and the complexity of weapons systems must be important determinants of command.

Command and control, therefore, are dependent on a number of factors — the vessels; equipment and armament; the social relations between crew, officers and admirals or admiralty; and the needs and dictates of the government — all of which can be analyzed using the environment-technology-culture triad.

The Anglo-American Tradition of Naval Command

Although all navies (and armies and air forces) are "commanded," they are not all commanded alike. Again, the differences are in the details. While it should seem axiomatic that navies stemming from a democratic tradition should practise a different form of at least higher-level command than very centralized totalitarian systems practise, such as in the Nazi German *Kriegsmarine* or more latterly the Soviet Navy, this factor has received little scholarly attention. A useful starting point for such analysis would be the notion that different societies will produce different navies, which ostensibly will employ different command styles.

One recent popular account edges into this territory by making the specific point that "maritime powers have always prevailed over land-based empires…revealing the way in which supremacy at sea freed thought and society itself."[16] Since the author's aim was to broadly distinguish the Anglo-American navies from those of the empires of continental Europe, he included no detailed assessment of different command styles, other than to infer an Anglo-American commander's greater scope for independence of action. Although hardly scientific, it is no great leap to postulate that the general principles can be extended to those navies that follow in the same tradition.[17] That relatively small grouping includes, besides Great Britain and the United States themselves, only Australia, New Zealand, arguably the Netherlands, and Canada.

An appreciation of the origins and development of the environmental, technological and cultural influences that make up what can be called the Anglo-American tradition of naval command is vital to understanding command in the Canadian Navy today. There are four significant ages in the evolution of that tradition, corresponding to the growth and development of the Royal Navy (RN) and, latterly, the United States Navy (USN) that sprang from it. The first comprised the steady evolutionary development of the sailing navy, from King Henry VIII to the start of the reign of Queen Victoria. The second was the Victorian era, involving so much radical change that no true pattern emerged. The third was the first half of the twentieth century, ending with the defeat of the Axis Powers in the Second World War. The fourth is the current era, from the end of the Second World War to the early twenty-first century, where change has been just as radical as in the Victorian era, but has taken place within a solid framework that has permitted it once more to be evolutionary.

The First Age (1545–1860): The Sailing Navy.[18] Throughout the age of sail, technology evolved at a steady if sometimes rapid pace; however, changes in technology were always limited by the physical environment in which it was employed — the sea. The British sailing navy evolved from a land-based concept of war at sea being fought and commanded by soldiers in floating castles assisted by professional mariners, to a concept of the warship commanded by officers who were both professional mariners and professional war fighters. This made a significant difference to the way in which the RN conducted war at sea, and a new naval culture evolved from the technical and command changes that took place during the age of sail. The changes made by Pepys and others to institutionalize the professionalism of the naval commander endure in most Western navies today. And the modern equivalents of these tests of professional competence, such as the Canadian Navy's rigorous sea-command qualification process, are a testament to the wisdom of those who laid down the foundations of naval professionalism.

As the RN grew in size and complexity during successive wars in the age of sail, new control methods were introduced to help senior naval commanders exercise command over their fleets and ships, especially when fighting fleet actions. A relatively centralized control structure evolved based upon various versions of the written *Fighting Instructions*, and these proved adequate for the RN's needs until the demands imposed by the truly global nature of the Napoleonic Wars and the unprecedented size of the RN led the Admiralty to endorse a new command and control framework made famous by Nelson and his "band of brothers." Nelson pioneered the concept of *shared intent* for large-scale RN actions. While, in the past, captains had always been given an indication of the Admiralty's intent for missions far from home, fleet actions had been closely controlled by the directions found in the *Fighting Instructions*. Nelson, however, introduced to the RN a concept, called *distributed leadership* in *Leadership in the CF*. For Nelson and his band of brothers this meant forging a great deal of implicit intent through various means so that the amount of explicit intent communicated through orders could be minimized. This allowed Nelson to employ a relatively decentralized command and control system that proved highly successful in the Napoleonic Wars. However, within ships, a relatively rigid hierarchy still existed in which opportunities for distributed leadership were limited. Likewise, emergent leadership was not evident outside the hierarchy except in mutinies. These cultural norms of command and leadership were to change dramatically in the second age of the Royal Navy.

The Second Age (1860–1906): Radical Change.[19] The second age of the RN can be characterized as one where unprecedented and rapid technological change had a major impact on the culture of the navy. Nevertheless, the pace and the nature of technological change were often determined by those whose careers had been made in the age of sail. The example of a ship's appearance taking precedence over proficiency in gunnery is an example of a reaction to technology that was not uncommon in this period.[20] But the demands for the effective use of naval power eventually forced change on the RN, requiring it to choose among the types of new technology available in order to modernize its fleet and to professionalize all members of the navy, not just the executive or deck officers. The granting of career status to the ratings and the creation of specialist officer classes had a long-term impact on the culture of the RN. Naval command during the *Pax Britannica* eschewed the risks associated with the distributed leadership style that was practised by Nelson and his band of brothers, and as communications technology improved, centralized control structures, like the *Fighting Instructions*, and a culture of caution once again limited the freedom of action of naval commanders.

As the nineteenth century and the *Pax Britannica* drew to a close, a new age was dawning for the RN that would be strongly influenced by the past — a past that was also to shape the culture of a child of the twentieth century, the new Canadian Navy.

The Third Age (1906–1945): The Two World Wars.[21] The first half of the twentieth century witnessed the paradox of some stability in major warship types, permeated nonetheless with the continued evolution of naval technology at a rapid pace, creating opportunities for some and obstacles for others. The changes in the RN that were influenced by technology could be seen at both the ship and the admiralty levels.

On the one hand, in the environment of larger ships, sailors became workers in floating factories, with their duties increasingly specialized compared to their forebears in the age of sail. Working in small groups, they became more and more isolated from their shipmates and focused on their specific tasks. On the other hand, the RN exploited new technology, especially in the area of anti-submarine warfare, as it became a large navy of small ships during the Second World War. This small-ship environment affected the work of crews in that there was less specialization, and sailors were less isolated from their shipmates.

The two work environments, large and small ship, created a dichotomy of command cultures in the RN of the first half of the twentieth century. In the major units of the fleet, officers used more authoritarian forms of leadership and more centralized command and control methods to exercise command over their large crews. In the small ships with their smaller complements, however, teamwork and co-operation, the hallmarks of the navy in the future, were the norm. The dichotomy also applied to leadership experience as officers who spent most of their careers in large ships tended to focus on specialist technical skills to the detriment of leadership roles, while officers who spent most of their careers in small ships, with earlier command opportunities, were less specialized but acquired more leadership experience. One could argue that in small ships, officers and petty officers needed to rely more on personal power (for example, expert, referent and connection, as described in *Leadership in the CF*) than on the more traditional position power (for example, reward, coercive, information and ecological) employed in large-ship command hierarchies.

Another characteristic of shipboard leadership and command culture in the first half of the twentieth century was the increased prominence of emergent leadership seen in the roles taken by specialist non-commissioned officers and ratings, especially the radar and Asdic operators. This trend was magnified by the culture of the small-ship environment, which valued the co-operative characteristics of expert teams as opposed to the more rigid hierarchies of the large-ship environment. The influx of civilians into the RN, particularly during the Second World War, also favoured the acceptance of emergent leadership by those with expert power, paralleling the same phenomenon in a British society in which leadership was becoming less hierarchical and less class based.

At the admiralty level, the increased communications capabilities developed in the twentieth century allowed flag officers ashore to institute increasingly centralized command measures based on their ability to exercise command through more sophisticated control structures. This reduced the effectiveness of the RN in the First World War when the new centralized control capabilities, like wireless telegraphy, reinforced the dominant, large-ship, cultural norm of deference to higher authority, in order to severely restrict the inclination of fleet commanders and captains to exercise their initiative.[22] During the Second World War, despite capabilities to exercise more centralized command and control due to improved communications technology, and with more concerns about communications security, captains exercised more initiative. This phenomenon was caused by the rise of officers from the small-ship navy into positions of senior command and by the environment in which the growing small-ship navy operated. These officers came from a culture that valued initiative, and they increasingly dominated RN command positions. The cultural shift was reinforced by an environment where the many small ships scattered on operations around the globe made centralized command difficult and made distributed leadership based on co-operation and co-ordination an effective way to conduct operations.

The joint and combined nature of the Second World War accelerated the trend that was begun in the First World War of the RN relying on increasingly larger and more sophisticated staffs to assist commanders ashore in the exercise of their command.

The Fourth Age (1945–present): The Modern Navy.[23] The senior command of the Royal Navy has had to adjust to these changed circumstances in unprecedented ways. Although the navy had partnered with foreign navies in the past, Britain had always been the dominant partner. Beginning in the Korean War and then through the Cold War, the RN had to adjust to the status of a member of an American-led coalition, progressively happy to follow American tactics and doctrine, and increasingly dependent on the Americans for their quantitative and qualitative lead in technology. So far, that leadership has continued into the post–Cold War era.

The real change for allied navies has been in the degree of civilian control. The entry of the US Navy into the Second World War meant that the RN had an equal partner with less experience, but American production soon made the USN the dominant partner in the Pacific and one not to be taken for granted in the European theatre. The retreat from empire by the British, the emergence of the USA and the USSR as the two superpowers, and the advent of the Cold War left the USN as the lead in the partnership. This meant a juggling of roles. Britain sought to maintain its leadership role in the Commonwealth and in its own foreign policy, while assuming a leadership role as a European member of NATO. It nonetheless had to acknowledge a secondary and conformist position to the USN in the global Cold War.

A new style of command, developed from the growing acceptance during the Second World War of emergent leadership styles, known as *command by negation* has been used to describe the phenomenon of increased responsibility being granted to technical experts, initially among warships' crews, but from there increasingly dispersed up the command chain, reaching its apex in the *Composite Warfare Commander* concept adopted by NATO in the 1980s.[24] In the late twentieth and early twenty-first centuries, the increased use of fully automated systems, with no humans in the decision loop, put another face on command by negation that needs to be explored more fully.

The historical development of the Anglo-American command style shows that the elements of the environment-technology-culture triad not only interact among themselves but also affect the nature of command and leadership over time.

Historical Determinants of Canadian Naval Command

Within the larger Anglo-American tradition, the experience of the Canadian Navy presents a persuasive case study in which each element of the environment-technology-culture triad has played a critical part in defining a distinctive naval command style.[25]

Despite attempts to define Canadian naval heritage as extending back to the *ancien régime*, there is no clear continuity of that continental tradition, and for all of its existence the modern service clearly has held itself to be part of the Anglo-American tradition. That is a point of no small import on several levels: from its start, the service accepted as given that it was an integral part of a winning tradition; its officers generally have practised a quite enlightened treatment of their sailors, in keeping with a system in which the harsh "Captain Bligh" leadership style was clearly atypical;[26] and it has maintained an unquestioning belief in "objective civilian control" as a core element of civil-military relations.[27]

The Canadian navy started out in 1910 as a virtual clone, culturally, of the Royal Navy. For the first four decades of its existence, since almost all training of officers and specialist ratings was conducted in Britain and much of their sea experience was with the RN, British naval cultural values were diffused throughout the new Canadian navy. Many of these values, such as mastering the naval profession, stood the Royal Canadian Navy (RCN) in good stead at the tactical level of command. However, at higher levels of command the lack of exposure to the workings of

admiralty and experience with higher-level staff work left the RCN
dependent on officers who were transferred on loan from the Royal Navy
to expand the RCN's strategic- and operational-level staffs during the
Second World War.

Having learned the consequences of technological backwardness during
the Second World War, the Canadian navy has since striven to remain at
the forefront of technological change, but it has had to manage this
change within a context of restricted budgets and manpower ceilings.
The navy's overriding concern has been how to maintain a degree of
independence from the dominant world power (formerly Britain and now
the United States), and yet achieve an operationally useful level of
interoperability with those forces while maintaining a distinctive
Canadian identity.

There is one significant respect in which the Canadian navy has always
differed (and can be expected to continue to differ) from its British and
American antecedents: size. Canada's navy has always been a small-ship
navy, and even though many of its senior officers had large-ship
experience with the RN, their Canadian command experience was with
small vessels. As described in the previous section, this smallness
favoured teamwork and co-operation and more reliance on personal
power (for example, expert, referent and connection) than on the more
traditional position power (for example, reward, coercive, information
and ecological) employed in large-ship command hierarchies. This may
be part of the reason that the Canadian naval culture prides itself on a
quite enlightened treatment of its sailors compared to some other navies.

The small size of the Canadian navy imbued its strategic command
culture with two characteristics: a magnified impact of commanders'
personalities on command culture, and confusion of administrative-
operational responsibility. The small number of officers eligible for senior
command and the influence of the most senior of them in selecting their
successors have exaggerated the influence of senior Canadian naval
officers compared to the senior officers of larger allied navies. And
whereas the larger American and British services have the depth to
establish an "admiralty" organization capable of accomplishing both
operations and administration functions, the small number of Canadian
officers available and qualified for staff work has led to a somewhat

artificial strategic-operational split between the Maritime Staff in Ottawa and the fleet commanders on each coast. It is telling that in the United States Navy the position of head of the navy is Chief of Naval Operations, while in Canada the position is styled Chief of the Naval Staff or Chief of the Maritime Staff.[28]

At the operational and tactical levels, Canadian naval commanders have attempted to compensate for the small size of their fleet through a combination of doctrinal innovation and technology. Typically restricted to possession of destroyer-size vessels, Canadian officers have tended to adapt them to novel uses more along the lines of light cruisers. In the late interwar years, tactics concentrated upon stealthy night attacks against heavier opponents, and despite their limited range the vessels voyaged far from home waters into the Caribbean and down the Pacific coast of Latin America. In the 1960s, there was the more famous "marriage" of a big anti-submarine warfare helicopter (the venerable Sea King) to the small deck of a destroyer. Among the lesser-known innovations of that same period was the Canadian lead in producing the first functional inter-ship, data-link, command information system in DATAR (the Digital Automatic Training and Remote System), a forerunner to Links 4 and 11. More recently, the widely dispersed formations for towed array operations required the acquisition of strategic communications systems, such as SATCOM, and associated inter-ship plotting systems.

Brought together in the 1990s, these technical developments had significant implications for the nature of command in the Canadian Navy. Previous notions of rigid command and control optimized for close-in anti-submarine warfare no longer were appropriate. At the ship level, individual commanders discovered a new independence, requiring greater emphasis on their initiative and technical competence. Operational-level commanders found their tactical horizons broadened significantly beyond the immediacy of close-in convoy escort.

Meanwhile, the culture of the Canadian navy evolved significantly also. The predominance of RN culture changed in the Second World War with the huge expansion of the RCN and a concomitant massive influx of civilians into the RCN's ranks. The experience of the war and the expansion of the RCN in the Cold War, after its immediate post–Second World War downsizing, combined to provide it with the critical mass of personnel necessary to justify creating a Canadian training establishment for junior officers and ratings. While many cultural traditions of the RN persisted in the Canadian Navy, the new Canadian training establishments resident in an evolving Canadian society, along with new roles for the navy, slowly but surely created a new uniquely Canadian naval culture. This culture diverged even further from RN culture in the last 30 years of the twentieth century with the introduction of French-language units and the acceptance of women, first in sea trades, then in command positions. The Canadian naval culture, then, is built on the keel of the professionalism of the Royal Navy, but its superstructure reflects Canadian values and experiences.

Notwithstanding the British base of Canadian naval culture, ever since the Second World War the Canadian navy has absorbed certain aspects of the US Navy's culture. Starting with communications procedures, technical systems, and living arrangements, through the policy of interoperability the Canadian navy has adopted more and more of the US Navy's doctrine and standard operating procedures. Therefore, from a combined point of view, the Canadian navy has achieved the "seamless operational integration at short notice" with the US and other allies mandated by the government.[29]

While the Anglo-American tradition is the foundation upon which Canadian naval command style is based, the relatively small size of the Canadian Navy compared to its American and British cousins has resulted in a unique Canadian naval command culture, shaped by Canadian culture and enabled by discerning choices in technology. One aspect of this uniqueness has been the Canadian Navy's ability to maintain "command parity" with its larger cousins in order to retain Canada's independence of action in naval operations. Without the ability to maintain a viable command and control framework, the Canadian Navy would have no option but to put its ships under command of larger coalition partners.

Command parity was first demonstrated by Rear-Admiral L.W. Murray in the Second World War, as Commander-in-Chief, Canadian Northwest Atlantic area. But it took the Korean War to refresh the lesson learned in the Second World War that to maximize the chances of achieving Canadian strategic and political aims, wherever possible, the principle should be respected that Canadian warships on foreign deployments should be under the operational command of a Canadian, as a recognizable national naval task group. This principle was used to inform the work of the Canadians involved in developing the *Concept of Maritime Operations* idea, which became the foundation of the *Task Group Concept*, as the basis for Cold War, fleet tactical employment. These command concepts introduced command by negation, a command style previously seen at the ship or low tactical level, to the *Composite Warfare Commander* at the operational level during the Cold War, where *Subordinate Warfare Commanders*, by virtue of their expert and information power, could display a type of emergent leadership within a pre-planned broad scope of action.

As the Cold War ended, Canadian naval commanders were rediscovering their earlier command parity, with possession of world-class C^2 systems leading to their assignment of the NATO subordinate warfare area responsibilities of anti-submarine warfare commander. On this basis, the Canadian task group commander in the 1991 Persian Gulf War became the only non-USN officer to hold a significant command appointment, overseeing the Coalition Logistics Force. The true novelty of the situation lay in the fact that it was exercised within an ad-hoc coalition structure as opposed to a formal alliance, a point underscored by the commander's redefinition of C^2 to mean *co-operation* and *co-ordination*.[30]

At the end of the twentieth and the beginning of the twenty-first centuries the Canadian Navy has continued to use its command parity capability to exercise high tactical level and low operational level command, for example, with multinational embargo operations around Haiti in 1993–1994, and with STANAVFORLANT operations in the Adriatic off former Yugoslavia. The modern expression of the Canadian Navy's command culture culminated in Operation Apollo, allowing the navy to adapt to the shift and expansion of its roles in the Arabian Sea in the winter and spring of 2003 to achieve the operational level of command in task force operations (CTF 151). This process was a classic illustration of the environment-technology-culture triad.

The Canadian Naval Command Style

The redefinition of C^2 in coalition operations to mean co-operation and co-ordination reflects the reality of command in the future where coalition operations may predominate. This new paradigm of co-opera-tion and co-ordination appears to emphasize leadership or influence behaviours among peers more than traditional concepts of command involving the exercise of authority over subordinates. Therefore, in coalition operations the concepts of emergent leadership and distributed leadership may be more useful than concepts of authority. In fact, one might see the high reputation that senior Canadian naval officers have earned in certain operational command positions as a type of emergent leadership based on three subclasses of personal power (that is, expert, referent, and connection), rather than position power.

One could argue that Canada's national culture with its traditions of bilingualism and multiculturalism, Canada's military culture with its history of alliance and UN operations, and Canadian naval culture based on operational and command competence, enlightened leadership and management techniques, and a judicious exploitation of available technology all combine to make the Canadian Navy's command style a model for coalition operations.

Without an in-depth analysis of navies around the world, which is beyond the scope of this chapter, it is not possible to answer the question posed in the introduction: is there something that can be designated a distinctly "naval" command style? Nothing uncovered in the course of research to date, however, contradicts the generally accepted preliminary conclusion that different societies produce different naval command styles. At its broadest, this suggests that navies stemming from a democratic tradition should practise a different form of at least higher-level command than do very centralized authoritarian systems.

It can be claimed with some confidence that the Canadian Navy is a legitimate progeny of the Anglo-American tradition of naval command. The following command characteristics help to define naval command styles in the Anglo-American tradition:

- **The Professional Standard of the Mariner.** One of the profession-al requirements of the naval commander is to meet or exceed the professional standard of the mariner. It may well be this requirement that is the primary cause of naval commanders, trained for command, possessing a ruthless determination to ensure that the ship is ready to move quickly at all times, and being able to make tough decisions without hesitation.

- **Competence, Confidence and Arrogance.** The command characteristic that is created by the rigorous command-qualification process used by the navies that share the Anglo-American tradition is one of competence, confidence and even a touch of arrogance.

- **Independence.** Throughout history, and to some extent even in today's networked world, naval commanders had to be prepared to make decisions that might have serious and wide-ranging consequences. Therefore, naval commanders operate in a culture that encourages and prizes independence. Despite the fact that Canadian Forces' Contingent Commanders are now provided with detailed instructions as to their responsibilities and authority, naval commanders without those instructions are still expected to act when they believe that it is necessary to do so.

- **Unique Officer Leadership Competencies.** In the Anglo-American tradition, junior officers undergo a long process of formal training and evaluation that is conducted and overseen principally by experienced officers. Officers in training receive relatively little instruction or mentoring from non-commissioned members (NCMs), unlike the army. Therefore, even quite junior command-qualified naval officers have skills that senior NCMs do not have.

- **Status and Aura of Command.** In addition to the power held by the commanding officer of a warship, the organization and the training system of the navy cause the naval commander to possess a significant status and aura of command.

- **Aggressive Leadership.** The naval commander understands the importance of knowing operations inside and out, not only to fight the ship but also to lead effectively. It is very likely that most naval commanders will appear to be aggressive and quick to make decisions during operations.

- **The Primacy of Training.** The naval commander will do anything in his or her power to obtain and to conduct all of the training necessary to bring the ship's company to the highest level of readiness possible and to keep it there. Commanders who do not will find their command appointments shortened and their prospects limited.

To these can be added unique Canadian attributes. Canada is conscious of its membership in the NATO alliance, the United Nations, the Commonwealth and other multilateral organizations. Rather than expecting others to adhere to its standards, it strives to maintain communications, both technical and social, with all other nations. At the same time, Canadian ships tend to be far more capable than most like-sized vessels of other navies. This leaves the Canadian Navy well placed, with its close relations with the USN and its historical relations with the RN — and its obligation to neither — to serve as a communications exchange between the navies of the world. In a commercial environment likened to a global village, the patchwork of coalition navies requires a medium of communication and co-ordination, a role for which Canadian naval commanders are well positioned.

Conclusions

The effectiveness of new naval C^2 systems and concepts will in large part be determined by how compatible they are with existing naval command styles. While technology can effect change in some dimensions of naval command styles, other dimensions will be resistant to change, often with good reason. The nature of operations at sea defines many aspects of naval command, and technical systems that take this fact into account will be more effective than those that do not. Likewise, naval culture, developed over centuries of war at sea, has many functional aspects that must be considered by those who design technical systems.

Naval command styles differ among nations, navies and commanders. The factors that influence naval command styles are many and varied, and yet all of them must be considered to understand the phenomenon of naval command. Canadian naval command styles are based on a culture that is founded on the professionalism of Canada's Navy. Professional expertise is acquired through long and rigorous training under the supervision and mentoring of experienced naval officers. At the tactical level, effective naval commanders are expected to employ a wide range of leadership behaviours. At the task group level, Canadian naval staffs have effected a judicious combination of technical decision-support with various personal attributes to create a unique style of command. This has led to a preference for Canadians to assume certain senior command appointments in recent coalition operations.

Canadian naval command styles, therefore, may vary in the details but collectively are unique in many ways. The small size of the Canadian Navy has imbued its command culture with a number of characteristics: a magnified impact of commanders' personality on command culture; the confusion of administrative-operational responsibility among the various headquarters levels; an appreciation for the discerning application of C^2 technology; and the innovative development of doctrine to foster command parity with larger allies. The result is a classic demonstration of the environment-technology-culture triad of naval command.

NOTES

1   This chapter is an overview of a more in-depth study undertaken for Defence Research and Development Canada (DRDC), Toronto, to describe and analyze naval command styles as part of a larger project investigating automated command advisory systems. The study, by Allan English, Richard Gimblett, Lynn Mason and Mervyn Berridge Sills, is "Command Styles in the Canadian Navy," Defence Research and Development (DRDC), Toronto, Contract Report CR 2005-096 (31 January 2005).

2   See Allan English, *Understanding Military Culture: A Canadian Perspective* (Montreal & Kingston: McGill-Queen's Univ. Press, 2004) chaps. 4, 5 and 6, for a detailed discussion of some of these issues in the Canadian and American military context. See also Allan English, Richard Gimblett and Howard Coombs, "Beware of Putting the Cart Before the Horse: Network Enabled Operations as a Canadian Approach to Transformation," DRDC Toronto, Contract Report CR 2005-212 (19 July 2005), for a discussion of networked operations.

3   For example see Oliver Warner, *Command at Sea: Great Fighting Admirals from Hawke to Nimitz* (New York: St Martin's, 1976), and Jack Sweetman, ed., *The Great Admirals: Command at Sea, 1587–1945* (Annapolis, MD: US Naval Institute Press, 1997).

4   The following is a selected alphabetical listing: Alan Easton, *50 North: An Atlantic Battleground* (Toronto: Ryerson, 1963; Markham, ON: Paperjacks, 1980); Michael Hadley, Rob Hubert and Fred W. Crickard, eds., *A Nation's Navy: In Quest of Canadian Naval Identity* (Montreal and Kingston: McGill-Queen's University Press, 1996); Peter T. Haydon, *The 1962 Cuban Missile Crisis: Canadian Involvement Reconsidered* (Toronto: CISS, 1993); Duncan E. Miller and Sharon Hobson, *The Persian Excursion: The Canadian Navy in the Gulf War* (Clementsport, NS: Canadian Peacekeeping Press, 1995); and William H. Pugsley, *Saints, Devils and Ordinary Seamen: Life on the Royal Canadian Navy's Lower Deck* (Toronto: Collins, 1945). The record will be only partially improved with the forthcoming survey of Chiefs of the Naval Staff and Commanders of Maritime Command by the CF Leadership Institute: Michael Whitby, Richard

Gimblett and Peter Haydon, eds., *The Admirals: Canada's Senior Naval Leadership in the Twentieth Century.*

5   Before unification Canada had three separate services: the RCN, the RCAF, and the Canadian Army. After unification in 1968, to emphasize the point that Canada no longer had three services, Department of National Defence (DND) bureaucrats coined the rather awkward term *environment*, based on the environments in which the sea, air, and land components of the CF operate, to describe these three components of the CF. Since there is only one military service in Canada today, the CF, official DND publications sometimes use the noun *environment* and the adjective *environmental* when referring to the sea, air, and land components of the CF. Nonetheless, the terms *Canadian Army, Navy* and *Air Force* are creeping back into official usage.   In this chapter the term *Canadian Navy* is used to refer to the post-unification maritime component of the CF, and the term environment is used to refer to the physical operating surroundings.

6   Italics in the original. DND, *Leadership in the Canadian Forces: Conceptual Foundations* (Kingston, ON: Canadian Defence Academy, 2005), 10.

7   See *Leadership in the Canadian Forces*, 8–11, for a more detailed discussion of these issues.

8   It should be stressed that the term *admiralty* has never been used in Canadian practice, but it is employed here to distinguish that higher strategic level from the ubiquitous use of the term *headquarters*.

9   Ross Pigeau and Carol McCann, "Re-conceptualizing Command and Control," *Canadian Military Journal* 3, no. 1 (Spring 2002), 53–63. See also their chapter in this book.

10   Carl H. Builder, *The Icarus Syndrome: The Role of Air Power Theory in the Evolution and Fate of the US Air Force* (London: Transaction Publishers, 1994), 5–7.

11   DND, *Leadmark: The Navy's Strategy for 2020* (Ottawa: Chief of the Maritime Staff, 2001); also at http://www.navy.dnd.ca/mspa_news/news_e.asp?id=11.

12   John Keegan, *The Mask of Command* (London: Jonathan Cape, 1987), 10 and *passim*.

13   The best openly available discussion of the Canadian process is J.Y. Forcier and David Hudock, "On an Even Keel: Warship Command Competency Beyond the Technical Skills," unpublished MA thesis, Royal Roads University (August 2000).

14   For example, lieutenant-commander is *capitaine de corvette*; commander is *capitaine de frégate*; and captain is *capitaine de vaisseau*.

15   See Michael Lewis, *A Social History of the Navy, 1793–1815* (London: Allen & Unwin, 1960), 85–6, and 270–80, for descriptions of the crew and their duties; idlers usually comprised less than 10 percent of the crew of about 800 on a first-rate ship of the line.

16   Peter Padfield, *Maritime Supremacy and the Opening of the Western Mind: Naval Campaigns that Shaped the Modern World, 1588–1782* (Woodstock, NY: Overlook, 1999), quote from jacket notes; and Peter Padfield, *Maritime Power and the Struggle for Freedom, 1788–1851* (London: John Murray, 2004). See also Colin Gray, *The Leverage of Sea Power: The Strategic Advantage of Navies in War* (New York: Free Press, 1992). However, neither of these authors explores how the nature of command is different.

17   Christopher Bell and Bruce Elleman, *Naval Mutinies of the Twentieth Century: An International Perspective* (London: Frank Cass, 2003), 264–76, in their concluding chapter, "Naval Mutinies in the Twentieth Century and Beyond," add validity to this presumption in noting the distinctions in the ways that democratic and totalitarian powers react to the idea of mutiny.

18   The dates signify the firing of the first broadside, 15 August 1545, to the launching of the steam-driven, screw-propelled, armoured, iron-hulled HMS *Warrior*. English et al., "Command Styles in the Canadian Navy," has a fuller discussion of this broad topic in section 3, "The Historical Development of the Anglo-American Naval Command Culture," 63–88.

19   Dating from HMS *Warrior*, the first armoured ironclad, to HMS *Dreadnought*, the first modern capital ship.

20   The US Navy's post–Civil War approach to technology is described as "a wholesale and deliberate policy of technical retrogression" by Robert L. O'Connell, *Of Arms and Men* (New York: Oxford University Press, 1989), 213.

21   Dating from HMS *Dreadnought* to the dropping of the atomic bomb.

22   Andrew Gordon, *The Rules of the Game: Jutland and British Naval Command* (London: John Murray, 1996), is a compelling examination of the development of these two different command styles in the Royal Navy through the nineteenth century, and their culmination at the Battle of Jutland in admirals Jellicoe and Beatty.

23   Dating from the dropping of the atomic bomb, through the Cold War, to the present day.

24   The Composite Warfare Commander was introduced to allow the effective management of increasingly complicated modern naval warfare. Under the general guidance of an overall "principal" commander, responsibility for each of the anti-air, anti-submarine, anti-surface and strike duties was sub-divided among "Subordinate Warfare Commanders" who would "command by negation" (that is, juniors are authorized to operate within a pre-planned, broad scope of action unless overridden by senior commanders).

25   It would be impracticable for the purposes of this chapter to recreate the entire history of the Canadian Navy. For a general discussion of the broader theme, see English et al., "Command Styles in the Canadian Navy," section 4, "Environmental, Technological and Cultural Determinants of Canadian Naval Command Styles – An Historical Perspective," 89–111.

26   N.A.M. Rodger, *The Wooden World: An Anatomy of the Georgian Navy* (Annapolis, MD: Naval Institute Press, 1986), 237–38, puts this myth to rest with the observation that "the violent seizure of a ship from her officers, on the high seas…[which] may be said to belong to the Cecil B. de Mille school of history…[was] virtually unknown in the Navy." For a discussion of the Canadian experience, see Richard H. Gimblett, "What the Mainguy Report Never Told Us,: The Tradition of Mutiny in the Royal Canadian Navy before 1949," *Canadian Military Journal* 1, no. 2 (Summer 2000), 87–94, also at http://www.journal.forces.gc.ca/engraph/Vol1/no2/pdf/85-92_e.pdf.

27   Samuel P. Huntington, *Soldier and the State: The Theory and Politics of Civil-Military Relations* (Cambridge, MA: Harvard University Press, 1957), viii. Although he wrote this seminal volume based primarily upon the experience of the US Army, it often is overlooked that he began his academic career as a student of the US Navy, with his "National Policy and the Transoceanic Navy," *US Naval Institute Proceedings* 80, no. 5 (May 1954).

28   The equivalent British term is *First Sea Lord of the Admiralty*. The origin of the title *Chief of the Naval Staff* is an interesting and early example of the integrationist tendencies of the Canadian Forces, dating back to the tenure of Walter Hose, who forced the adoption of the term in 1928 as a means of establishing his parity with successive Chiefs of the [Army] General Staff, who constantly attempted to absorb the naval service.

29   DND, *Shaping the Future of the Canadian Forces: A Strategy for 2020* (June 1999), http://www.cds.forces.gc.ca/pubs/strategy2k/intro_e.asp.

30   Richard Gimblett, "MIF or MNF? The Dilemma of the 'Lesser' Navies in the Gulf War Coalition," in Hadley et al., *A Nation's Navy*, 193.

CHAPTER 3

COMMAND IN A COMPLEX BATTLESPACE

Colonel Christian Rousseau

> People in this world look at things mistakenly, and think that what they do not understand must be the void. This is not the true void. It is bewilderment.
>
> Miyamoto Musashi [1]

The elusive search for certainty in military decision making has been amply documented by the likes of Carl von Clausewitz and Martin van Creveld: certainty about the state and intentions of the enemy's forces; certainty about the environment in which the war is fought; and certainty about the state, intentions and activities of one's own forces. Every advance in the sophistication of control systems[2] reflects this race between the demand for information and the systems' ability to deliver it. And until very recently, as van Creveld points out, taming uncertainty still proved a chimera:

> Taken as a whole, present-day military forces, for all the imposing array of electronic gadgetry at their disposal, give no evidence whatsoever of being one whit more capable of dealing with the information needed for the command process than were their predecessors a century or even a millennium ago...their ability to approach certainty has not improved to any marked extent.[3]

But that was in 1985, ages ago in the fast-growing field of information technology. If we are to believe the present-day enthusiasts,[4] it seems that 100 percent or perfect battlespace visibility (BV)[5] is closer at hand than ever. The powerful analogy of putting the commander back on his horse is used to describe the phenomenon:

> The battle commander no longer needs to overlook the battlefield; he no longer needs to be in the vicinity of the battle;

he no longer needs to be adjacent to the arena of battle; he no longer needs to be in even the same global hemisphere of the battle. The image of the 19th century general astride his horse surveying the battle on a vast plain below him has been replaced by that of the 21st century general viewing a cluster of video screens and digital maps that portray battle changes in real time.[6]

Achieving perfect BV is no small feat. "Many intelligence reports in war are contradictory; even more are false, and most are uncertain."[7] And "in practice, the incoming information is of inconsistent value: 99 percent of it is likely to disappear without a trace, whereas the remaining 1 percent may have a profound effect on operations — though whether this means that the 1 percent would be of value even without the 99 percent is a different question altogether."[8]

If we suspend disbelief momentarily and assume that not only is perfect BV achievable but its interface could be designed in such a way as to eliminate the risk of information overload, would this considerable expenditure in resources represent a significant gain in the commander's ability to make the right decision at the right time?

This chapter will argue that even perfect BV would only bring marginal value to the commander and that war at the operational level will remain a complex endeavour, requiring exacting decision-making skills and coping strategies to make sense of the complexity.

To present this argument we will first investigate *complexity theory* and show its relevance to the Joint Force Commander (JFC) and his or her environment. The second part will focus on some of the latest research done in the field of decision making in complex environments and will contrast findings with the situation of a JFC. In the last part, we will distil the enablers to operating in war's complex environment from the insights gained by the realization of its chaotic nature.

Complex Systems

Everything in war is simple, but the simplest thing is difficult.
Carl vonClausewitz[9]

Terms like *complexity, chaos* and *non-linearity** have cropped up in our day-to-day vocabulary, signalling a pervasive use of the theories behind them to explain our world. The reign of the predictable Newtonian world has given way to the flux of the capricious world of chaos. But what are chaos and complexity theories, and what do they have to do with commanders or the theory of warfare? We will answer these two questions in turn.

There is an important link between the theories of complexity and chaos. Succinctly, *chaos* is the study of how simple systems can generate complicated behaviour, while *complexity* is the study of how complicated systems can generate simple behaviour.[10] A familiarity with both concepts is important to our understanding of warfare. Let us look at chaos first to set the scene for the introduction of complexity theory.

Although it owes its birth to mathematics, chaos is now a multi-disciplinary science. The great French mathematician Henri Poincaré first noticed the idea that many simple non-linear deterministic systems can behave in an apparently unpredictable and chaotic manner.[11] Other early pioneering work in the field of chaotic dynamics was found in the mathematical literature; however, the importance of chaos was not fully appreciated until the widespread availability of digital computers for numerical simulations and the demonstration of chaos in various physical systems. This realization has broad implications for many fields of science, and it is only within the past decade or so that the field has undergone explosive growth. It has been found that the ideas of chaos have been very fruitful in such diverse disciplines as biology, economics, chemistry, engineering, fluid mechanics, and physics, to name a few.[12]

The thing for the layperson to recognize is that chaos is not randomness; rather, the phenomenon of chaos is a very sensitive dependence of the outcome of a process, in a deterministic system, on the tiny details of what happened earlier, the initial conditions. When chaos is present, it amplifies indeterminacy.[13] But if all non-linear systems were completely indeterminate, not much would come out of their study. Complexity theory, for its part, deals with the study of systems that exhibit unpre-

---

* The terms complexity, chaos and non-linearity are used in this chapter in their mathematical (or scientific) sense rather than their day-to-day meaning of complicated, unorganized and non-contiguous. See the glossary at the end of this chapter for specific definitions.

dictable, but within bounds, self-organizing behaviour. One of the defining features of complex systems is a property known as emergence in which the global behaviour of the system is qualitatively different from the behaviour of the parts. No amount of knowledge of the behaviour of the parts would allow one to predict the behaviour of the whole.[14]

The other point to appreciate is that when one is dealing with systems, interactions are the norm. Action in one area will invariably have more than one effect. We are dealing with a system when (1) units or elements are interconnected so that changes in some elements or their relations produce changes in other parts of the system, and (2) the entire system exhibits properties and behaviours that are different from those of the parts. As a result, systems often display non-linear relationships, outcomes cannot be understood by adding together the units or their relations, and many of the results of actions are unintended. Complexities can appear even in what would seem to be simple and deterministic situations. In a system, the chains of consequences extend over time and many areas, and the effects of action are always multiple. Doctors call the undesired impact of medications "side effects." Although the language is misleading — there are no criteria other than our desires that determine which effects are "main" and which are "side" — the point reminds us that disturbing a system will produce several changes.[15]

Finally, it should be evident from the above discussion that further complexities are introduced when we look at the interactions that occur between strategies when actors consciously react to others and anticipate what they think others will do.[16]

Now that we have a basic understanding of what chaos and complexity theories are, we will look at the Joint Force Commander and his operating environment to show that they constitute indeed a chaotic and complex system. For a system to be considered complex,* it must be deterministic, and its interactions must induce non-linearity and be, within bounds, self-organizing. If we have all three conditions (deterministic, non-linear, and pattern-forming self-organization), then it can be considered a complex system.

---

* While recognizing that this is not a universal quality of complexity, in the instances we deal with in this chapter, chaotic behaviour is a precursor to complexity. Therefore, in the remainder of the chapter, to lighten the text, the term *complex* will be used to mean "chaotic and complex."

That the constituting elements of war are deterministic there can be little doubt. When a carrier battle group sails, it does not randomly travel around the world's oceans; when a fighter squadron flies on a mission, it does not drop ordnance arbitrarily; and when an armoured division attempts to take an enemy position, advancing erratically does not serve its purpose.[17] Just the fact that such groupings exist and have shown to be potent systems to control and inflict violence signifies that there is a link between cause and effect. We are not dealing with a stochastic environment.

That interactions in war induce non-linearity is well documented. From the nursery rhyme told to stress the importance of taking care of small problems to forestall bigger ones —

> For the want of a nail, the shoe was lost;
> For the want of the shoe, the horse was lost;
> For the want of the horse, the rider was lost;
> For the want of a rider, the battle was lost;
> For the want of the battle, the kingdom was lost;
> All for the want of a nail....[18]

— to the learned studies of Clausewitz —

> in war...countless minor incidents — the kind you can never really foresee — combine to lower the general level of performance, so that one always falls far short of the intended goal.... This tremendous friction, which cannot, as in mechanics, be reduced to a few points, is everywhere in contact with chance, and brings about effects that cannot be measured, just because they are largely due to chance....[19]

— or Helmuth von Moltke's remark that "no operation plan extends with any certainty beyond the first encounter with the main body of the enemy,"[20] practitioners, theorists and even popular culture testify to the futility of predicting results based on initial conditions, considering their sensitivity to seemingly benign perturbations. Non-linear outcomes are the hallmark of war; the latter's nature cannot be captured in one place but emerges from the collective behaviour of all the individual agents in the system interacting locally in response to local conditions and partial information. In this respect, decentralization is not merely one choice of command and control; it is the basic nature of war.[21] Furthermore, non-linearity also comes from the sophistication of the organization itself.[22] It will be no surprise to anyone who has worked in any sizeable headquarters, particularly joint and combined ones, that its internal operation can be chaotic. More precisely, it can be turbulent and weakly chaotic, exhibiting features of self-organized criticality.[23]

The pattern-forming self-organization aspect of warfare can be glimpsed from studying its history, or more precisely from the fact that it is possible and worthwhile to use history to enhance our understanding of warfare. If warfare were not pattern forming, the introduction of new technology that changes the balance of interactions, on one side or both, would bring unrecognizable new dynamics in the system. Yet when we look at the functions of war over time (the requirement to Sense, Shield, Act, Sustain and Command), they have been impervious to technological change.[24] The emergence of principles of war is also a sign of pattern-forming self-organization. If there were no pattern, only non-linearity, we could not affirm that concentration of force is worth pursuing and that selection and maintenance of the aim is an enabler to success, and we could conclude that the wisdom of keeping a reserve is an anachronism from the nineteenth century.[25]

The last remaining task to show that warfare is a complex environment is to discern whether we are truly dealing with a complex system or simply a metaphor. Social sciences are often subjective, and complexity theory has become trendy.[26] There have been many attempts in the past to transpose concepts from the "hard" sciences to the "soft" ones with mixed results.[27] Although there is little doubt that the three prerequisites of complex systems are met, to prove that it is actually valid, our new theory would have to make verifiable predictions that are not explicable by other theories of warfare.[28] Nevertheless, whether warfare is actually complex, or simply behaves like a complex system, is too fine a point for the purposes of this chapter. In either case, the commander has to deal with pattern-forming unpredictability.

It is therefore clear that war, the environment in which a JFC operates, is a complex system where knowing the physical component of the situation is only part of the solution. Non-linear dynamics suggests that war is uncertain in a deeply fundamental way. Uncertainty is not merely an initial environment condition that can be reduced by gathering information and displaying it on a computer screen. It is not that we currently lack the technology to gather enough information but will someday have the capability. Rather, uncertainty is a natural and unavoidable product of the dynamic of war; action in war generates uncertainty.[29] How can we help a commander deal with these complex systems? How much help would a perfect battlespace visibility system bring? After investigating how the human mind deals with complexity, we will be in a good position to answer these questions.

Decision Making in a Complex System

> This difficulty of *accurate recognition* constitutes one of the most serious sources of friction in war, by making things appear entirely different from what one had expected.
>
> Carl von Clausewitz[30]

Despite our seemingly advanced cognitive skills, it appears that evolution allowed human beings to develop a tendency to deal with issues on an ad-hoc basis. The early problems we had to deal with when the task at hand was to gather firewood, drive a herd of horses into a canyon or build a trap for a mammoth were problems of the moment and usually had no significance beyond themselves. The need to see a problem embedded in the context of other problems rarely arose. For the modern day JFC, however, thinking in terms of complex systems is the rule, not the exception. Do our habits of thought measure up to the demands of thinking in systems? What errors are we prone to when we have to take side effects and long-term repercussions into account?[31] These questions will be answered by looking at some of the latest research done in the field of decision making in complex environments. We will first investigate the apparent limitations of the human mind and the consequent type of recurring decision errors in complex environments. That will set the stage for us to explore strategies for successful decision making. We will then be in a position to contrast these findings with the situation of a JFC.

At the root of our difficulty in dealing with complex systems is our poor ability to deal with variable patterns in time.[32] The fact that spatial configurations can be perceived in their entirety while temporal ones cannot may well explain why we are far more able to recognize, and deal with, arrangements in space than in time. We are constantly presented with whole spatial configurations and readily think in such terms. We know, for example, that to determine whether a parking lot is crowded we need to look at more than one or two spaces. By contrast, we often overlook time configurations and treat successive steps in a temporal development as individual events. For example, as enrolment rises each year, the members of a school board may add first one room, then another, onto an existing schoolhouse because they fail to see the development in time that will make an additional schoolhouse necessary.[33] Even when we think in terms of time configurations, our intuition is very limited; so when we have to cope with systems that do not operate in accordance with very simple temporal patterns, like the one given here, we run into major difficulties. There are two types of relatively straightforward temporal patterns that create undue difficulties: non-linear growth or shrinkage (the magic of compounded interest), and developments that show changes of direction like oscillations or sudden reversals.[34]

Limited temporal intuition is evident in our propensity to "oversteer" when action and reaction are not linked by instantaneous feedback. At the helm of the proverbial oil tanker, the uninitiated (non-expert) will keep turning the wheel because the ship appears non-responsive. Once it starts to turn, we realize that we have overdone it and have to compensate the other way.

> This tendency to "oversteer" is characteristic of human interaction with dynamic systems. We let ourselves be guided not by development within the system, that is, by time *differentials* between sequential stages, but by the *situation* at each stage. We regulate the *situation* and not the *process*, with the result that the inherent behaviour of the system and our attempts at steering it combine to carry it beyond the desired mark.[35]

Unfortunately, limited temporal intuition and tendency to oversteer do not appear to be our only flaws. Dealing with uncertainty seems to be another vulnerability. Dietrich Dörner, an authority on cognitive behaviour, found that decision makers who are uncomfortable with complexity and unfamiliar with a situation are often plagued with uncertainty[36] and so tend to

- act without proper analysis of the situation,

- fail to anticipate side effects and long-term repercussions,

- assume that the absence of immediately obvious negative effects means that correct measures have been taken,

- let over-involvement in "projects" blind them to emerging needs and changes in the situation, and

- be prone to cynical reactions when encountering failure.[37]

They also tend to miss the big picture and be swamped in trying to deal with the problem of the moment: "One reason they deal with partial problems in isolation is their preoccupation with the immediate goals.... At the moment, we don't have other problems, so why think about them? Or to put it better still, why think that we should think about them?"[38]

Experts, for their part, deal with complexity within their field in stride[39] but remain vulnerable to uncertainty. Gary Klein, an authority on naturalistic decision making, found that experts familiar with the complexity of a particular situation make three types of error: error due to lack of experience; error due to lack of information; and what he calls the *de minimus*[40] error, an error of mental simulation where the decision maker notices the signs of a problem but explains it away. He or she finds a reason not to take seriously each piece of evidence that warns of an anomaly.[41]

The foregoing makes clear that decision making in a complex environment does not come naturally. Cognitive psychology scientists have documented strategies to effective decision making in such environments, which differ whether or not the decision maker is an expert in the field where the decisions are required. But before going into the strategies, we need to look at an emerging truth that seems to hold regardless of expertise levels.

In successfully dealing with complex environments, it appears that cognitive ability is not the main indicator and that the usual battery of psychological tests is useless in predicting participant behaviour. One would assume that "intelligence" would determine behaviour in complex situations, since complicated planning — formulating and carrying out decisions — presumably places demands on what psychology has traditionally labelled *intelligence*. But Dörner has found that there is no significant correlation between scores on IQ tests and performance in his problem-solving experiments. It seems that a better predictor of participant success is the individual's capacity to tolerate uncertainty. [42]

When we want to operate within a complex and dynamic system, we have to know not only what its current status is but also what its status will be or could be in the future, and we have to know how certain actions we take will influence the situation. For this we need *structural knowledge*, knowledge of how the variables in the system are related and how they influence one another.[43] As we will see when discussing recognition-primed decision making, experts have developed an intuition[44] for this, but laypersons must hypothesize the links, test the hypotheses and keep in mind the possibility that their model is probably wrong.[45] We will look at this approach to successful decision making first.

The decision-making strategy proposed by Dörner is very similar to what we have come to understand as the estimate of the situation, or in collaborative planning terms, the operational planning process:

> *Defining goals* is the first step in dealing with a complex problem, for it is not immediately obvious in every situation what it is we really want to achieve…. *Developing a model and gathering information* follow the statement of goals….[46] *Prognosis and extrapolation* is the third step … to assess not only the status quo but also developments likely to follow from the current situation…. Our next step is to *consider measures to achieve our goals*. What should we do? Should we do anything at all?… *Decisions* follow planning…. *Action* follows decisions. Plans must be translated into reality. This, too, is a difficult enterprise, one that calls for constant self-observation and critique. Is what I expected to happen actually happening? Were the premises for

my actions correct, or do I have to go back to an earlier phase of the planning process and retool?[47]

Even if this kind of planning process serves us well in unfamiliar complex situations, it is rather slow and cumbersome and not always practicable under extreme time pressure. Fortunately it appears that shortcuts are possible for situations where the decision maker has a sound structural knowledge (implicit or explicit) of the system he or she is dealing with. As Dörner explains,

> [For the human mind] complexity* is not an objective factor but a subjective one. Take, for example, the everyday activity of driving a car. For a beginner, this is a complex business. He must attend to many variables at once, and that makes driving in a city a hair-raising experience for him. For an experienced driver, on the other hand, this situation poses no problem at all. The main difference between these two individuals is that the experienced driver reacts to many "supersignals." For her, a traffic situation is not made up of a multitude of elements that must be interpreted individually. It is a gestalt,[48] just as the face of an acquaintance, instead of being a multitude of contours, surfaces, and color variations, is a "face." Supersignals reduce complexity, collapsing a number of features into one. Consequently, complexity must be understood in terms of a specific individual and his supply of supersignals. We learn supersignals from experience.[49]

Gary Klein studied experts in their natural settings with ample structural knowledge and a good grasp of supersignals. He found that "the commanders could come up with a good course of action from the start...."

> Even when faced with a complex situation, the commanders could see it as familiar and know how to react. The commander's secret was that their experience let them see a situation, even a non-routine one, as an example of a prototype, so they knew the typical course of action right away. Their experience let them identify a reasonable reaction as the first one they considered, so

* In this case, the term *complexity* is not used strictly in its mathematical sense. Dörner uses it to mean "the difficulty for the human mind to deal with a given level of complexity."

they did not bother thinking of others. They were not being perverse. They were being skillful. We now call this strategy *recognition-primed decision making.*[50]

Recognition-primed decision making (RPD) fuses two processes: the way decision makers size up the situation to recognize which course of action makes sense (called *pattern recognition*), and the way they evaluate that course of action by imagining it (called *mental simulation*). The RPD strategy matches the following pattern: The decision makers recognize the situation as typical and familiar and proceed to take action. They understand what type of *goals* make sense (so the priorities are set), which *cues* are important (so there is not an overload of information), what to *expect* next (so they can prepare themselves and notice surprises), and the *typical ways of responding* in a given situation. By recognizing a situation as typical, they also recognize a *course of action* (COA) likely to succeed. They do not compare COAs. They wargame (mental simulation) the first plausible COA and use it as is, adjust it if need be, or reject it if it will not do the job. They do not attempt to find the best plan; they are after the first plan that they know will work, thereby achieving great economies of time and mental resources.[51] After the decision has been made, experts monitor developments and rely on their expectancies as one safeguard. If they read a situation correctly, the expectancies should match the events. If they are wrong, they can quickly use their experience to notice anomalies and change their plan dynamically.[52]

This strategy of decision making has its limitations, however, and cannot serve in all situations. Significant structural knowledge (mainly implicit) is required, and there is a relatively low limit to how complex a situation can be before it overwhelms our mental capabilities to simulate it. Because of our short-term memory limitations, the simulation is limited to a maximum of three moving parts and has to do its job in no more than six steps. We have to assemble the simulation within these constraints. Furthermore, if the variables interact with each other, the job of visualizing the program in action becomes even more difficult, and so we search for a way to keep the transitions flowing smoothly by building a simulation that has as few interactions as possible.[53] Moreover, with our difficulty in dealing intuitively with all but the simplest temporal pattern, mental simulation and RPD will not help in circumstances where complex temporal configurations are at play.

We have seen what researchers found to be our wanting qualities when it came to successful decision making in complex environments. They have highlighted the errors of our ways and suggested strategies to overcome them. We will now illustrate the applicability of their theories by demonstrating their usefulness at explaining the Joint Force Commander's situation.

We recognize in Klein's decision-making modus operandi a pattern applicable to the commander, possessing what Clausewitz has called *coup d'oeil*.[54] Similarly, we can see the close parallel between the Operational Planning Process (OPP) and Dörner's guidelines for decision making in unfamiliar, complex situations. His highlighting of the difficulty of executing a plan echoes Thomas Czerwinski's comments on the same subject: "increased complexity has kept pace with heightened competency; ... command-by-plan inherently fights the disorderly nature of war as much as the adversary."[55]

In the same way we can easily find examples of the three types of errors reported by Klein,[56] and an informed reading of Cohen and Gooch's book on military misfortunes reveals that lack of structural knowledge is at the root of the three kinds of failure they report.[57] This deficiency is clearly at play when Thomas remarks that "information superiority allowed NATO to know almost everything about the battlefield [in the Kosovo conflict], but NATO analysts didn't always understand everything they thought they knew."[58] And in the incident Mandeles describes:

> Generally, senior commanders find it difficult during combat both to distinguish outputs from outcomes and to discover outcomes. In fact, the inability to discern outcomes (damage to specific enemy capabilities) is usually the reason senior commanders focus strongly on outputs, such as sortie rates.[59]

Finally, the concept that tolerance towards uncertainty is a better predictor of success than sheer intellect can be corroborated by the words of William Tecumseh Sherman telling a subordinate what made Ulysses Grant his superior in the art of war:

> Wilson, I'm a damned sight smarter than Grant; I know more about organization, supply and administration and about

everything else than he does; but I'll tell you where he beats me and where he beats the world. He don't care a damn for what the enemy does out of his sight but it scares me like hell. I'm more nervous than he is. I am much more likely to change my orders or to countermarch my command than he is. He uses such information as he has according to his best judgment; he issues his orders and does his level best to carry them out without much reference to what is going on about him....[60]

It is clear from the above that the theories described in this part of the chapter are applicable to the JFC and that making decisions in the complex dynamic system characteristic of warfare is no simple matter for him or her. On their own, our cognitive abilities do not seem robust enough to deal directly with high-level complexity and require coping strategies in the form of pattern matching and decision-making schemes. This makes evolutionary sense since complexity often self-organizes in patterns. Armed with this understanding we can now investigate what enablers to decision making are available and how to integrate them into the JFC's world.

Enablers to Effective Decision Making in Complex Systems

Even amidst the tumult and the clamour of battle, in all its confusion, he [the expert at battle] cannot be confused.

Sun Tzu[61]

Enablers to decision making for commanders are not a new idea. From Machiavelli to Czerwinski, authors have attempted to investigate, collate and enunciate principles, schemes and philosophies to assist commanders in arriving at proper decisions. The aim here is not to confirm, deny or replace the work of those authors or give an exhaustive list of dos and don'ts. Rather, it is to look at some of the insights gained by realizing that the JFC deals in an unpredictable, yet within bounds, self-organizing, complex environment. These insights are grouped in two broad categories: those related to the commander and those that affect command within the organization. At the root of these insights are two primordial principles: the first is to recognize that *time is the scarce commodity* (an organization has to be able to match the rate of change in its environment); and the second is to recognize that *people are the key asset of any organization* (people are the adaptive element of organizations, and learning and innovation come only from human cognition).[62]

The first insight related to commanders is their comfort level in a chaotic environment. We have already noted that a capacity to tolerate uncertainty was a better predictor of success than straight cognitive ability. Being at ease with chaos would permit the commander to profit from it rather than waste energy fighting it.[63] Maybe there is more to the German tongue-in-cheek adage about the classification of officers than we chose to believe in the past:

> I divide my officers into four classes as follows: the clever, the industrious, the lazy, and the stupid. Each officer possesses at least two of these qualities. Those who are clever and industrious I appoint to the General Staff. Use can under certain circumstances be made of those who are stupid and lazy. The man who is clever and lazy qualifies for the highest leadership posts. He has the requisite nerves and the mental clarity for difficult decisions. But whoever is stupid and industrious must be got rid of, for he is too dangerous.[64]

We are not advocating reserving the JFC position to the laziest officer around. However, selecting people who are at ease with chaos and nurturing that talent would be to our advantage. In the words of Dörner:

> An individual's reality model can be right or wrong, complete or incomplete. As a rule it will be both incomplete and wrong, and one would do well to keep that probability in mind.... The ability to make allowances for incomplete and incorrect information and hypothesis is an important requirement for dealing with complex situations. This ability does not appear to come naturally, however. One must therefore learn to cultivate it.[65]

The second issue related to successful decision making in individuals, and therefore applicable to the JFC, is what is referred to as *operative intelligence* or *metacognition*.[66] In dealing with complex problems, we cannot handle in the same way all the different situations we encounter. By understanding their own cognitive limitations, experts can choose problem-solving strategies that maximize their strengths and minimize their weaknesses. For example, we noted earlier in this chapter our poor ability to deal with variable patterns in time. It appears from experimentation that using graphs to convert "time" into "space" helps people comprehend temporal configurations.[67] An understanding of our limitation allows us to devise strategies to deal with it.

Four components of metacognition seem most important: memory limitations, having the big picture, self-critiques and strategy selection. Experts, by being sensitive to their own memory limitations and how they affect their mental simulation capabilities, adopt subtle procedures to avoid the difficulty and factor in their level of alertness, their ability to sustain concentration, and so forth. When it comes to the big picture, experts not only see it, they can detect when they are starting to lose it. Rather than wait until they have become hopelessly confused, experts sense any slippage early and make the necessary adaptations. The self-critique ability in experts comes from their performance being less variable than that of novices; they can more easily notice when they do a poor job and can usually figure out why in order to make corrections. Using these abilities, experts can think about their own thinking to change their analytic strategies.[68]

Where this metacognition comes from and how do we impart it to the potential JFCs? Experience seems to be the answer. Dörner tells us:

> Geniuses are geniuses by birth, whereas the wise gain their wisdom through experience. And it seems to me that the ability to deal with problems *in the most appropriate way* is the hallmark of wisdom rather than genius.[69]

Or, in the words of Clausewitz describing the remedy to his familiar friction: "Is there any lubricant that will reduce this abrasion? Only one, and a commander and his army will not always have it readily: combat experience."[70] Direct experience is the area most fertile for providing improvement in decision-making performance in the complex environment we have described.[71] Considering the dearth of combat experience at the operational level in our militaries, training becomes the vehicle of choice to gain those habits that will make us better decision makers. The rest of the insights related to the commander deal with how we prepare him or her for the job. Most of them deal with the professional development triad of education, training and experience, and weigh heavily on the side of experience.

Let us first look at the value of concentrating on the educational route for improving the decision maker's ability to deal with complex situations. Dörner explains the results of such an approach:

> The training gave them what I would call "verbal intelligence" in the field of solving complex problems. Equipped with lots of shiny new concepts, they were able to *talk* about their thinking, their actions, and the problems they were facing. This gain of eloquence left no mark at all on their performance, however. Other investigators report a similar gap between verbal intelligence and performance intelligence and distinguish between "explicit" and "implicit" knowledge. The ability to talk about something does not necessarily reflect an ability to deal with it in reality.[72]

When a training approach is applied (that is, lesson closely followed by practice) to teach formal methods of analysis, it proves a hindrance to rapid decision making. Klein explains:

> We do not make someone an expert through training in formal methods of analysis. Quite the contrary is true, in fact: we run the risk of slowing the development of skills. If the purpose is to train people in time-pressured decision making, we might require that the trainee make rapid responses rather than ponder all the implications.[73]

Because expertise depends on perceptual skills, and perceptual learning takes many cases to develop, you rarely get someone to jump a skill level by teaching more facts and rules. In natural settings, perceptual learning grows with the accumulation of experience. Powerful training methods will not grow instant experts; the most we can expect from them is to make training more efficient.[74] So, left with frequent practice as the sole reliable contributor to improved decision making in complex environments, and a scarcity of combat experience, how can we prepare our decision makers for the challenge?

The best approach to replicate the required experience is a robust, yet accessible, simulation program: a program filled with exercises and realistic scenarios, where teaching takes a backseat to practice, allowing the person a chance to size up numerous situations very quickly. A good simulation can sometimes provide more training value than direct experience; it lets you stop the action, back up to see what went on, and cram many trials together so a person can develop a sense of typicality.[75]

The next insight, dealing with maximizing the value of each experiential event, reinforces the importance of the *After Action Review* (AAR) process as we know it. While teaching is of little value in developing the ability to make decisions in a complex environment, it appears that conscious self-reflection makes a difference. Subjects who were asked to reflect on their own thought process after each iteration of a problem-solving experiment performed much better than a control group who were asked to do something unrelated.[76] This forced foray into metacognition made them better problem solvers. Self-reflection can be enhanced by the presence of an expert observer who, having witnessed how the participant planned and acted and having noted his or her errors and their determinants, assists the participant in his or her reflection through carefully prepared follow-up sessions.[77]

Two related words of caution on these last two insights are necessary. First, the customary warning on the fidelity of the simulation used to replicate the reality of the experiential event: if the patterns in the simulation do not match the ones in reality, the subject develops the wrong intuition. The second caution, more sombre, indicates that, despite our best efforts at simulation, we may never truly develop expertise in our subject area. According to Klein:

> We will not build up real expertise when: The domain is dynamic, we have to predict human behavior, we have less chance for feedback, the task does not have enough repetition to build a sense of typicality, [or] we have fewer trials. Under these conditions, we should be cautious about assuming that experience translates into expertise. In these sorts of domains, experience would give us smooth routines, showing that we had been doing the job for a while. Yet our expertise might not go much

beyond these surface routines; we would not have a chance to develop reliable expertise.[78]

This might explain why peacetime generals often get sacked at the beginning of a war. They have acquired experience but have had no chance to develop expertise. If we are condemned to collecting experience in our field rather than expertise, the best advice for preparing for war may be that offered by Mandeles: "In war a commander needs a set of organizations that will learn while they execute their missions. What those organizations can practice in peacetime is not so much precisely what to do in war, but how to learn quickly what to do."[79]

We will therefore now turn our attention to the insights that deal with command within the organization, that is, insights into the thinking apparatus of the joint force or, more precisely, the ways in which planning is carried out, decisions are made or delegated, and intent is communicated.

Our first realization is that the concept of metacognition can be applied to the apparatus of the joint force itself to create the right groupings, structure and information flow to maximize its strengths and minimize its weaknesses.[80] When it comes to planning, we have remarked earlier at the similarity of the decision-making strategy proposed by Dörner to the Operational Planning Process and noted that although this serves us well in unfamiliar, complex situations, it is slow and cumbersome and difficult to apply under extreme time pressure. Klein's model of recognition-primed decision making, on the other hand, exploits the experience of the decision maker to produce rapid reaction, but it is limited to relatively familiar situations. In the OPP, the symbiosis between the two is meant to be the commander's planning guidance, where having sized up the situation, he or she decides on the goal and directs the COAs to be investigated. The staff then attempts to produce the plan that would make those COAs work. This takes advantage of the pattern-recognition skills of the most experienced member of the force, the joint force commander, and channels the energy of the staff's brainpower to dealing with the complexity that each COA represents. Unfortunately, the reality differs from the theory. Commanders, perhaps intimidated by the credentials of their "Ninja Team,"[81] tend to let planners come up with a COA, and then the commanders, realizing that it does not meet their understanding of the situation, incrementally adjust it to their liking during the information and decision briefs.[82] This may be a reflection of the teaching method in our staff colleges where directing staff, acting as commanders, ask their student planners to come up with the commander's planning guidance, ostensibly to give them a better opportunity to read in the problem. It is clear that delegating mission analysis and COA identification to the staff is the wrong approach; not only does it waste the time and cognitive energy of the staff, but it marginalizes the expertise of the commander who ultimately makes the decision.

Turning our attention to decision making and delegating, that is, command philosophy, we have been slow in the West to recognize that complexity demands the mission command approach. Even now, whenever technology floats the mirage of complete visibility of the battlespace, we let ourselves be tempted by the allure of more control. Unless the complete visibility promised also includes complete structural information (and it cannot),[83] mission command remains the only viable alternative. Using Dörner's words again:

> In very complex and quickly changing situations the most reasonable strategy is to plan only in rough outline and to delegate as many decisions as possible to subordinates. These subordinates will need considerable independence and a thorough understanding of the overall plan. Such strategies require a "redundancy of potential command," that is, many individuals who are all capable of carrying out leadership tasks within the context of general directives.[84]

Next, having sized up the situation, planned a response, made or delegated the decision, the thinking apparatus of the joint force has to communicate that intent to those who will implement it. The first enabler here, most familiar to military organization, is team building. Working with people who understand the culture, the task and anything that we are trying to accomplish allows them to "read our minds" and fill in the unspecified details.[85]

With implicit intent established, we need to deal with explicit intent. The notion of telling subordinates not only what to do, but why they must do it, is again a relatively new concept in Anglo-Saxon military heritage. The primary function of communicating intent is to allow better improvisation. Once we accept that in a complex system we cannot think of all the contingencies in advance and that we have to resort to mission command, giving the reasoning behind a task will allow subordinates to be creative. They will make adjustments to the plan based on the conditions in the field that the higher-level headquarters cannot know about. They will recognize opportunities that no one expected, and find ways to jury-rig solutions when the plan runs into trouble. Explicit intent should be clear enough for them to set and revise priorities and to decide when to grasp an opportunity and when to let it go.[86]

Commander's intent is already part of the orders format used by the joint force where intent, scheme of manoeuvre, main effort and end-state are given. Klein's list is more exhaustive and includes information that we normally only publish internally to the headquarters in the form of the Commander's Planning Guidance or the Chief of Staff's Planning Directive. Klein states:

> There are seven types of information that a person could present to help the people receiving the request to understand what to do: (1) the purpose of the task (the higher-level goals); (2) the objective of the task (an image of the desired outcome); (3) the sequence of steps in the plan; (4) the rationale for the plan; (5) the key decisions that may have to be made; (6) antigoals (unwanted outcomes); (7) constraints and other considerations.[87]

Although most of this information finds its way into the orders format, it might be worthwhile to re-examine our intent paragraph to make sure it meets all of our needs.

Some of the insights presented above, gained from the realization that the Joint Force Commander deals in an unpredictable, yet within bounds, self-organizing, complex environment, have already been adopted by Western militaries. The philosophy of mission command, the construct of the operational planning process, the idea of communicating intent, and the widespread use of after-action reviews are indicators that we have come to realize we have to learn to live with complexity. On the other hand, the practice of appointing commanders who are comfortable with chaos and the concept of metacognition is recognized but not yet ingrained in our culture. Frequent exposure, through simulation, to the realities of decision making in complex environments, rather than training in formal methods of analysis, would go a long way to ingraining this practice in our culture. Not only will the application of these insights save precious time in the decision-action cycle, but tackling the challenge of chaos with a focus on developing people will put us in good stead in our endeavour to become a truly learning organization.

Conclusion

> Having heard what can be gained from my assessment, shape a strategic advantage (*shih*) from them to strengthen our position.
>
> Sun Tzu[88]

War, the environment in which a Joint Force Commander operates, is a complex system where knowing the physical component of the situation, that is, battlespace visibility, is only part of the solution. Interactions, even deterministic ones, make war uncertain in a deeply fundamental way. Considering our cognitive limitations, making decisions in the complex dynamic systems that characterize warfare is no simple matter. Coping strategies in the form of pattern matching and decision-making schemes are required to make sense of the complexity. Some of these strategies and enablers, like the philosophy of mission command, have already been adopted by Western militaries. Others, like the wholesale acceptance of naturalistic decision-making methods, have yet to make inroads into most military thinking. Frequent exposure to the realities of decision making in complex environments, through simulation or otherwise, needs to figure predominantly in our professional development scheme of commanders.

Selection and development are two complementary approaches we can take to ensure our commanders thrive on chaos rather than fight it. When it comes to selection, we already spend significant resources testing the cognitive ability and motor-skill potential of candidates for specific military occupations (MOCs). Testing for comfort level with chaos, or aptitude to develop it, would be advantageous in our selection of candidates in "operator MOCs" that are liable to provide commanders. Then, after the ascertainment of a capacity to tolerate chaos, adapting our professional development triad to actually develop that capacity would be the next step. In this regard, experience that engenders expertise needs to figure prominently. Deployed operational experience, instrumented field exercises, and computer-simulated exercises, all supported by comprehensive after-action reviews, are the activities of significant value here. In the same vein, the focus on deliberate planning that characterizes our staff colleges has to be counterbalanced by the incorporation of naturalistic decision-making methods in the curriculum. It is not enough to learn to plan. Learning to make time-sensitive decisions during execution must be taught and practised if our military education institutions are to be worthy of the title Command and Staff Colleges.

Perfect battlespace visibility is a significant step forward for the information gatherers and managers in the headquarters, that is, the staff. To the commander, however, BV will not reduce the chaotic nature of the environment. It has very limited utility in reducing the difficulty of decision making. And BV, even when it is perfect, deals with the current point in time and is, therefore, only a small part of the picture the commander needs to consider to make decisions within the complex system of war. Furthermore, its near-perfect quality, giving the impression of clarity and finality, may lead the commander to concentrate on the space configuration of the situation rather than the more difficult temporal one. So, even with perfect BV, commanders need to step back and reflect and exploit their intuition, mental simulation and other sources of power to truly appreciate the situation and arrive at decisions based on variables far more subtle than those that can be captured on a computer screen.

Glossary[89]

chaos. Effectively unpredictable long-time behaviour arising in a deterministic dynamical system because of sensitivity to initial conditions. It must be emphasized that a deterministic dynamical system is perfectly predictable given perfect knowledge of the initial condition, and is in practice always predictable in the short term. The key to long-term unpredictability is a property known as sensitivity to initial conditions.

complexity. Complex systems are non-linear systems characterized by collective properties associated with the system as a whole that are different from the characteristic behaviours of the constituent parts.

criticality. A point at which system properties change suddenly.

deterministic. Dynamical systems are *deterministic* if there is a unique consequent to every state, and *stochastic* or *random* if there is more than one consequent chosen from some probability distribution (the "perfect" coin toss has two consequents with equal probability for each initial state; it is not deterministic). Most of non-linear science deals with deterministic systems.

non-linear. In algebra, we define linearity in terms of functions that have the property $f(x+y) = f(x)+f(y)$ and $f(ax) = af(x)$. In other words, linearity involves two propositions: (1) changes in system output are proportional to changes in input; and (2) system outputs corresponding to the sum of two inputs are equal to the sum of the outputs arising from individual inputs. Non-linear is defined as the negation of linear. This means that the result f may be out of proportion to the input x or y; that is, a small input may have an unpredictably large output like the proverbial butterfly flapping its wing causing a hurricane on the other side of the world.

pattern-forming self-organization. Systems where structure appears without explicit pressure or involvement from outside the system. In other words, the constraints on form (that is, organization) are internal to the system, resulting from the interactions among the components and usually independent of the physical nature of those components.

self-organized criticality. The ability of a system to evolve in such a way as to approach a critical point and then maintain itself at that point.

stochastic or random. Systems where there is more than one consequent to every state chosen from some probability distribution (the "perfect" coin toss has two consequents with equal probability for each initial state; it is not deterministic).

NOTES

1   Miyamoto Musashi, *A Book of Five Rings*, trans. Victor Harris (Woodstock, NY: Overlook Press, 1974), 95.

2   The Pigeau-McCann concept of control, made up of structures and process, devised in part to reduce uncertainty, is used in this chapter. See Ross Pigeau and Carol McCann, "Re-conceptualizing Command and Control," *Canadian Military Journal* 3, no. 1 (Spring 2002), 54.

3   Martin van Creveld, *Command in War* (Cambridge, MA: Harvard University Press, 1985), 265–66.

4   A typical example is Arnold Beichman, "Revolution in Warfare Trenches," *Washington Times*, 31 January 1996, 17; originally referred to in John D. Hall, "Decision-Making in the Information Age: Moving Beyond the MDMP," *Field Artillery Journal* (September–October 2000), 28.

5   For the purpose of this chapter, I will define perfect battlespace visibility as the ability for the commander to see, on demand, everything in the battlespace. I deliberately use the term *battlespace visibility* rather than *situational awareness*. Although our doctrine appears to have a simplistic view of situational awareness (SA), virtually equating it to battlespace visibility, the two are not synonymous. The naïve doctrinal definition is: "Quite simply, if you can answer the following questions accurately, 'Where am I?' 'Where are the good guys?' 'Where are the bad guys and what are they doing?' 'What is that?' and 'What am I supposed to be doing?' then you have situational awareness." See Canada, Department of National Defence, *Land Force Command*, B-GL-300-003/FP-000 (dated 21 July 1996), 115. Researchers in the field have advanced the concept significantly beyond this simplistic view. Mica R. Endsley, a leading figure in the study of the concept of situation awareness, defines it as "the perception of the elements in the environment with a volume of time and space, the comprehension of their meaning and the projection of their status in the near future." She recognizes three levels of SA: *Level 1, Perception*, of elements in current situation; *Level 2, Comprehension*, of current situation; and *Level 3, Projection*, of future status. The first level, which deals with perception of cues, is fundamental. Without basic perception of important information, the odds of forming a correct picture of the situation are extremely low. Comprehension, for its part, encompasses how people integrate multiple pieces of information and determine their relevance to achieve an understanding of the situation. To differentiate Levels 1 and 2, she employs the analogy of having a high level of reading comprehension as compared to just reading words. The highest level of SA, Level 3, implies the ability to forecast future situation events (including their implication) and dynamics. Ability at this level allows for timely decision making. This approach places SA as the main precursor to decision making, which supports the argument of this chapter: higher-level SA happens in the commander's mind. The ability to translate the symbols on the computer screen (presumably Level 1) into Level 3 is what this chapter is about. See Mica R Endsley, "Theoretical Underpinnings of Situation Awareness: A Critical Review" in *Situation Awareness Analysis and Measurement*, eds. M.R Endsley and D.J. Garland (Mahwah, NJ: Lawrence Erlbaum Associates, 2000), 3–4.

6   Robert K Ackerman, "Operation Enduring Freedom Redefines Warfare" in Signal Tribute: The Fight for Freedom, *Signal Magazine* 57, no. 1 (September 2002), 3.

7   Carl von Clausewitz, *On War*, eds. and trans. Michael Howard and Peter Paret (Toronto: Random House of Canada, 1993), 136.

8   Van Creveld, *Command in War*, 7.

9   Clausewitz, *On War*, 138.

10   See the Frequently Asked Questions section of the University of Colorado Web site at http://amath.colorado.edu/faculty/jdm/faq-[2].html.

11   Non-linear: In algebra, we define linearity in terms of functions that have the property $f(x+y) = f(x)+f(y)$ and $f(ax) = af(x)$. In other words, linearity involves two propositions: (1) changes in system output are proportional to changes in input and (2) system outputs corresponding to the sum of two inputs are equal to the sum of the outputs arising from individual inputs. *Non-linear* is defined as the negation of linear. This means that the result f may be out of proportion to the input x or y; that is, a

small input may have an unpredictably large output, like the proverbial butterfly flapping its wings and causing a hurricane on the other side of the world.

Deterministic: Dynamical systems are "deterministic" if there is a unique consequent to every state, and "stochastic" or "random" if there is more than one consequent chosen from some probability distribution. (The "perfect" coin toss has two consequents with equal probability for each initial state; it is not deterministic). Most of non-linear science deals with deterministic systems.

Chaos: Effectively unpredictable long-time behaviour arising in a deterministic dynamical system because of sensitivity to initial conditions. It must be emphasized that a deterministic dynamical system is perfectly predictable given perfect knowledge of the initial condition, and is in practice always predictable in the short term. The key to long-term unpredictability is a property known as sensitivity to initial conditions. See the Frequently Asked Questions section of the University of Colorado Web site.

12   This part on the evolution of the science of chaos borrows heavily from the University of Maryland Web site at http://www-chaos.umd.edu/.

13   Murray Gell-Mann, "The Simple and the Complex," in *Complexity, Global Politics, and National Security*, eds. David S. Alberts and Thomas J. Czerwinski (Washington, DC: National Defense University, 1997), 15.

14   John F. Schmitt, "Command and (out of) Control: The Military Implication of Complexity Theory," in *Complexity, Global Politics, and National Security*, eds. Alberts and Czerwinski, 233–35.

15   Robert Jarvis, "Complex Systems: The Role of Interactions," in *Complexity, Global Politics, and National Security*, eds. Alberts and Czerwinski, 46–48.

16   Ibid., 5.

17   The constituting elements used here and their attendant level of aggregation are for illustrative purposes only. We could have used any other levels of aggregation; for example, at the most basic elements of warfare, if we know with enough precision the skill level of a sniper, her mental and physical state, the distance and angle to the target, the weather conditions and other obstructions to visibility, and the characteristics of the weapon used, we can determine if she will hit the target. It is not a random coin toss. For a discussion on the deterministic aspects of the technological elements of war see Martin van Creveld, *Technology and War: From 2000 B.C. to the Present* (New York: Free Press, 1991), 314–15.

18   Normally attributed to Benjamin Franklin's *Poor Richard's Almanack* (1758), even if the present-day saying is somewhat different from the original: "For the want of a nail, the shoe was lost; for the want of a shoe the horse was lost; and for the want of a horse the rider was lost, being overtaken and slain by the enemy, all for the want of care about a horseshoe nail." See http://usinfo.state.gov/usa/infousa/facts /loa/bf1758.htm.

19   Clausewitz, *On War*, 138–39.

20   Count Helmuth Karl Bernard von Moltke cited in Peter G. Tsouras, *Warrior's Words: A Quotation Book: From Sesostris III to Schwarzkopf, 1871 BC to AD 1991* (London: Arms and Armour Press, 1992), 61.

21   Schmitt, "Command and (out of) Control," 232.

22   Cohen and Gooch clearly recognize this: "The kinds of misfortunes we have discussed in this book are not, however, the product of malevolent chance. Neither are they the sole responsibility of any single individual, not even the military commander. Instead, each is the consequence of the inherent fragility of an entire organization." Eliot A. Cohen and John Gooch, *Military Misfortunes: The Anatomy of Failure in War* (New York: The Free Press, 1990), 243. Mandeles also deals with this issue: "The paradox of modern military command at the theatre level is that, although it is the responsibility of one officer, it must be exercised within a set of complex organizations." Mark D. Mandeles et al., *Managing Command and Control in the Persian Gulf War* (Westport, CT: Praeger, 1996), 6.

23   Steven R. Mann, "The Reaction to Chaos," in *Complexity, Global Politics, and National Security*, eds. Alberts and Czerwinski, 148.

24   Van Creveld, *Technology and War*, 314.

25   Clausewitz, *On War*, 247.

26   Mann, "The Reaction to Chaos," 138.

27   The concept of *centre of gravity* comes to mind.

28   The conditions enumerated here are necessary but not sufficient to conclude that the theory fits the facts. To prove the sufficient criteria is beyond the scope of this chapter.

29   Schmitt, "Command and (out of) Control," 236–7.

30   Clausewitz, *On War*, 137.

31   Dietrich Dörner, *The Logic of Failure: Recognizing and Avoiding Error in Complex Situations* (New York: Metropolitan Books, 1996), 5–6.

32   There are more factors at play here that Dörner identifies — "the slowness of our thinking and the small amount of information we can process at any one time, our tendency to protect our sense of our competence, the limited inflow capacity of our memory, and our tendency to focus only on immediately pressing problems" — as the causes of the mistakes we make dealing with complex systems. For the sake of brevity, I have chosen to highlight only the temporal feature here, and other aspects are noted in passing. See Dörner, *The Logic of Failure*, 190.

33   Dörner, *The Logic of Failure*, 109.

34   Ibid., 107–52.

35   Ibid., 30.

36   Klein defines uncertainty as "doubt that threatens to block action." Key pieces of information are missing, unreliable, ambiguous, inconsistent or too complex to interpret, and as a result a decision maker will be reluctant to act. Klein sees four sources of uncertainty: (1) missing information (information is unavailable — it has not been received, or it has been received but cannot be located when needed); (2) unreliable information (the credibility of the source is low or is perceived to be low even if the information is highly accurate); (3) ambiguous or conflicting information (there is more than one reasonable way to interpret the information); and (4) complex information (it is difficult to integrate the different facets of the data). Gary Klein, *Source of Power: How People Make Decisions* (Massachusetts: MIT Press, 1999), 276–7.

37   Dörner, *The Logic of Failure*, 18.

38   Ibid., 87–8.

39   Experts appear to have an overall sense of what is happening in a situation: an ability to judge prototypicality. Where novices may be confused by all the data elements, experts see the big picture and they appear to be less likely to fall victim to information overload. See Klein, *Source of Power*, 152.

40   A *de minimus* explanation (coined by Perrow in his 1984 book, *Normal Accidents: Living with High-Risk Technologies*) is one that tries to minimize inconsistencies. The operator forms an explanation and then proceeds to explain away disconfirming evidence. See Klein, *Source of Power*, 66.

41   Klein, *Source of Power*, 274.

42   Dörner, *The Logic of Failure*, 27.

43   Ibid., 41.

44   Dörner calls intuition the totality of implicit assumptions in an individual's mind — assumptions about the simple or complex links and the one-way or reciprocal influences between variables. Dörner, *The Logic of Failure*, 41. Similarly, Klein notes that "intuition depends on the use of experience to recognize key patterns that indicate the dynamics of the situation." He adds, however, that because patterns can be subtle, people often cannot describe what they noticed or how they judged a situation as typical or atypical. Therefore, intuition has a strange reputation. Skilled decision makers know that they can depend on their intuition, but at the same time they may feel uncomfortable trusting a source of power that seems so accidental. Klein, *Source of Power*, 31.

45   Dörner, *The Logic of Failure*, 21–4.

46   Dörner goes on to explain that we need, of course, to do more with information than simply gather it. We need to arrange it into an overall picture, a model for the reality with which we are dealing. Formless collections of data about random aspects of a situation merely add to the situation's impenetrability and are no aid to decision making. We need a cohesive picture that lets us determine what is impor-

tant and what is unimportant, what belongs together and what does not — in short, that tells us what our information *means*. This kind of "structural knowledge" will allow us to find order in apparent chaos. Dörner, *The Logic of Failure*, 44–5.

47   Dörner, *The Logic of Failure*, 43–6.

48   The Gestalt school emphasizes perceptual approaches to thought, rather than treating thought as calculating ways of manipulating symbols. Gestalt theory is a broadly interdisciplinary general theory that provides a framework for a wide variety of psychological phenomena, processes and applications. Human beings are viewed as open systems in active interaction with their environment. The primacy of the phenomenal — recognizing and taking seriously the human world of experience as the only immediately given reality, and not simply discussing it away — is a fundamental assertion of Gestalt theory. It is the interaction of the individual and the situation in the sense of a dynamic field that determines experience and behaviour and not only drives external stimuli or static personality traits. See the Society for Gestalt Theory and its Applications Web site at http://www.gestalttheory.net/gtax1.html#kap2.

49   Dörner, *The Logic of Failure*, 39.

50   Klein, *Source of Power*, 17.

51   Ibid., 24–6.

52   Ibid., 35.

53   Ibid., 52–3.

54   Clausewitz, *On War*, 118.

55   Thomas J. Czerwinski, "Command and Control at the Crossroads," *Parameters* 26, no. 3 (Autumn 1996), 124.

56   Examples that easily come to mind are the Iraqis in the Gulf War for lack of applicable experience; the Canadians at Dieppe for lack of information; and the failure to foresee the attack on the World Trade Center as a *de minimus* error.

57   "Failure to learn, failure to anticipate, and failure to adapt." Cohen and Gooch, *Military Misfortunes*, 26.

58   Timothy L. Thomas, "Kosovo and the Current Myth of Information Superiority," *Parameters* 30, no. 1 (Spring 2000), 14.

59   Mandeles et al., *Managing Command and Control in the Persian Gulf War*, 5.

60   From Lloyd Lewis, *Sherman: Fighting Prophet* (New York: Harcourt Brace, 1932), 424, cited in Cohen and Gooch, *Military Misfortunes,* 244.

61   Sun Tzu, *The Art of Warfare*, trans. Roger Ames (Toronto: Random House, 1993) 120.

62   Robert R. Maxfield, "Complexity and Organization Management," in *Complexity, Global Politics, and National Security*, eds. Alberts and Czerwinski, 183–4.

63   Complex systems have the property that many competing behaviours are possible and the system tends to alternate among them. In fact, the ability of a complex system to access many different states, combined with its sensitivity, offers great flexibility in manipulating the system's dynamics to select a desired behaviour. By understanding dynamically how some of the complex features arise, we show that it is indeed possible to control a complex system's behaviour. See Leon Pond and Celso Grebogi, "Controlling Complexity" (College Park, MD: Institute for Plasma Research, University of Maryland, 1995), abstract, http://www-chaos.umd.edu/.

64   Attributed to General Kurt von Hammerstein Equord circa 1933. From Tsouras, *Warrior's Words: A Quotation Book*, 297.

65   Dörner, *The Logic of Failure*, 42.

66   A term coined by Klein, taken to mean "thinking about thinking" or seeing inside your own thought process.

67   Ibid., 143.

68   Klein, *Source of Power*, 158–9.

69   Dörner, *The Logic of Failure*, 193.

70   Clausewitz, *On War*, 141.

71   Klein, *Source of Power,* 42.

72   Dörner, *The Logic of Failure*, 196.

73   Klein, *Source of Power*, 30.

74   Ibid., 287.

75   Ibid., 42–3.

76   Dörner, *The Logic of Failure*, 195.

77   Ibid., 196–99.

78   Klein, *Source of Power*, 282.

79   Mandeles et al., *Managing Command and Control in the Persian Gulf War*, 6.

80   Pigeau and McCann certainly see this as a responsibility of "Creative Command." See "Re-conceptualizing Command and Control," 55.

81   *Ninja Team* is the informal name given to the core group of planners, often graduates of the School of Advanced Military Studies, in an army or joint headquarters.

82   General Schwarzkopf's first version of the plan for a ground attack in the Gulf is the typical example. His instruction to his planners was, "Assume a ground attack will follow an air campaign. I want you to study the enemy dispositions and the terrain and tell me the best way to drive Iraq out of Kuwait given the forces we have available," only to come up with his own course of action later. See Norman H. Schwarzkopf, *It Doesn't Take a Hero* (New York: Bantam Books, 1992), 354, 362.

83   No matter how precisely you measure the initial condition in these systems, your prediction of its subsequent motion goes radically wrong after a short time. Typically, the predictability horizon grows only logarithmically with the precision of measurement. Thus, for each increase in precision by a factor of 10, say, you may only be able to predict two more time units. See the Frequently Asked Questions section of the University of Colorado Web site at http://amath.colorado.edu/faculty/jdm/faq-[2].html.

84   Dörner, *The Logic of Failure*, 161.

85   Klein, *Source of Power*, 219. Pigeau and McCann refer to this as *implicit intent*. See Ross Pigeau and Carol McCann, *Re-defining Command and Control* (Toronto: Defence and Civil Institute of Environmental Medicine, 1998), 5–6, and their chapter in this book.

86   Klein, *Source of Power*, 223.

87   Ibid., 225.

88   Sun Tzu, *The Art of Warfare*, 104.

89   Taken from the Frequently Asked Questions sections of the University of Colorado Web site at http://amath.colorado.edu/faculty/jdm/faq-[2].html and of the Self-organized Systems Web site at http://www.calresco.org/sos/sosfaq.htm#1.1.

CHAPTER 4

ESTABLISHING COMMON INTENT:
THE KEY TO CO-ORDINATED MILITARY ACTION

Ross Pigeau and Carol McCann

In democracies the role of the military is articulated in government laws, edicts and orders that also define the boundaries within which their militaries must function. Considering the lethal capability possessed by militaries, it is not surprising that democratic societies have established elaborate rules, regulations and procedures for ensuring that militaries operate within the bounds of society's laws, edicts and orders. Rules, regulations and procedures are a society's explicit mechanism for controlling the behaviour of its military; in other words, they are society's explicit mechanism for exercising legal authority over its military. But as important as a well-articulated military role is, and as important as the rules, regulations and procedures for society's control of the military are, these factors are secondary to society's expectation that its military will *interpret* its role in society correctly and that it will interpret the *spirit* rather than the letter of these rules, regulations and procedures. Society expects its military to understand the myriad connotations implicit in the need for a military in the first place. Society expects that during difficult military operations, where the situation may be ambiguous, ill defined and uncertain, its military will correctly infer both the nature of the desirable response as well as its magnitude. In short, militaries must understand — and act upon — societal intent; if they do not, militaries risk fulfilling their missions in a manner inconsistent with societal expectations.[1]

Correctly interpreting an aim, purpose or objective — that is, correctly inferring intent — is a fundamental concept in military thought. Military doctrinal literature is rife with terms like *commander's intent, intent statements* and *enemy intent.*[2] Fulfilling societal intent, as described above, may be a military's first and most fundamental intent priority, but it is only one of many. For example, the smooth functioning of a military organization, particularly during operations, depends upon its members correctly inferring not only the commander's intent but also one another's intent, especially in unanticipated situations for which plans may not have been prepared. In fact, intent is such a profound concept, one with such rich and far-reaching implications, that we have made it key to our definition of command and control (C^2). In 1996, we defined C^2 as "the establishment of common intent to achieve co-ordinated action,"[3] and in 2000, we suggested a number of mechanisms for its propagation.[4]

The aim of this chapter is to explore the concept of intent and to expand our ideas on how intent can be shared among individuals — particularly among military members — to achieve *common intent*. Furthermore, we wish to argue that common intent is a military's primary means for achieving co-ordinated action, especially if it adopts such operational philosophies as mission command, effects-based operations or network-enabled operations. We will build upon our previous work on common intent, extending it into a more complete conceptual framework suitable for research and, more importantly, suitable for provoking military thought and discussion.

Co-ordinated Action

The establishment of common intent in C^2 is not an end in itself but a means to an end: specifically, to co-ordinate action in military operations. The requirement for co-ordinated action is, we assume, uncontested among most militaries. The extensive physical environments within which militaries must operate; the number of belligerents, civilians and coalition members with whom they must interact; the mountains of supplies required to sustain military operations; and the pressures for timely and rapid action as well as the serious consequences of that action — all of these, and more, dictate an absolute requirement for co-ordinated action.

We define co-ordinated action as the proper arrangement of resources and effort, both in time and in space, to harmonize intended mission effects. Furthermore, we hypothesize that there are two classes of approaches used by militaries for achieving co-ordinated action: (1) explicitly defining structures, communication protocols, and instructions for controlling the behaviour of military members in the service of mission objectives;

and (2) allowing uncontrolled, spontaneous behaviour to emerge that, hopefully, will be consistent with mission objectives.

Class 1 Approaches: Explicit Control. To a large extent, the first class of approaches for co-ordinating action is a paraphrase of our definition of military control: those structures and processes devised by command to enable it and to manage risk.[5] Achieving co-ordinated action through the efficient, explicit control of resources is a time-honoured method for accomplishing missions. One example of this approach is the operational planning process (OPP), a planning procedure that is extensively taught in military training.[6] Another is the sophisticated electronic systems that have been developed over the years to manage and control the distribution and sharing of intelligence and tactical information, as well as to facilitate voice communications. Military doctrine itself is a control mechanism wherein high-level guidance for action is given. Indeed, there are countless control structures and processes that exist to co-ordinate military action.

In our previous work, we have discussed the importance of control structures and processes in military organizations, asserting that "a good control system will marshal and co-ordinate available resources in a systematic and ordered way, with appropriate checks and balances, in order to efficiently accomplish mission objectives with as little uncertainty as possible."[7] We have also discussed the limitations of control.[8] For example, the unbridled creation of elaborate control structures and processes can lead to an *over-control* situation that inhibits initiative, creativity and prudent risk-taking behaviour among military members. Furthermore, because control mechanisms (for example, weapon systems, rules of engagement, standard operating procedures) are notoriously brittle — that is, they generalize poorly beyond the conditions for which they were developed — they are, therefore, less useful in unpredictable military situations. Lastly, all control comes at a logistical price. At some point the care and feeding of control structures and processes that grow unchecked will exceed the benefits they deliver (for example, the "tooth-to-tail" ratio becomes too small). The dangers of over-control aside, however, this class of approaches for achieving co-ordinated action has proven to be extremely beneficial to militaries.

Class 2 Approaches: Spontaneous Emergent Behaviour. Rather than impose control externally, Class 2 approaches advocate letting co-ordinated action emerge spontaneously, with little or no explicit direction. The idea is that independent entities interacting in complex environments will, over time, display emergent behaviour that is consistent with co-ordinated action.

Two objections could be raised against this approach. First, spontaneous emergent behaviour may be a haphazard phenomenon at best, making it of limited value. Second, even if spontaneous emergent (and co-ordinated) behaviour is possible, the importance of achieving complex military operations safely and effectively, especially in politically and emotionally charged circumstances, makes this class of mechanism too uncertain and hazardous to be of value.

Interestingly, the first objection is unsupported by scientific evidence. Much of what we observe in nature can be classified as spontaneously emerging order.[9] Nicolis and Prigogine have established that during high-energy throughput, dissipative structures (that is, order) can emerge in a system where none existed before.[10] Two specific physical examples are the hexagonal convection currents (called Bénard cells) that arise when water is gently heated, and the change from laminar to turbulent flow in water as it exits from a tap under increasing pressure. Both examples illustrate the emergence of higher forms of order that allow more efficient transfer of energy (that is, heat in the first case and water volume in the second). More spectacular forms of emergent order occur in biological systems, such as viruses, bacteria, unicellular organisms and multicellular organisms (for example, flora and fauna). All these are examples of lower-level forms of order (for example, genetic material) emerging into higher forms of order (for example, cell bodies, entire organisms) by converting considerable amounts of energy (for example, light, food) while, at the same time, interacting in complex environments.

Finally, still higher forms of co-ordinated order emerge when large numbers of individual organisms interact. Among organisms this "social" behaviour does not even require a common "goal" or common "purpose." For instance, stigmergy, a science developed by the entomologist Grassé in 1959,[11] describes the complex emergent behaviour of "social" insects like ants, termites, wasps and bees. Individual insects in insect communities respond only to the vagaries of the local environment (including neighbouring insects) without any "knowledge" of their larger purpose.[12] As very large numbers of simple insects carry out their individually prescribed genetic behaviour in evolutionarily suitable environments, "social" order emerges that benefits the survivability of the entire community. The science of stigmergy has been used to study and explain such co-ordinated actions as flocking behaviour in birds, schooling behaviour in fish and pack behaviour among wolves.

The first objection to using emergent behaviour as a mechanism for achieving co-ordinated action is, therefore, invalid. But the second objection may yet have merit. Though there are numerous examples of spontaneous, co-ordinated behaviour in nature, it may be unwise for militaries to rely on this phenomenon during critical operational circumstances. After all, spontaneous order in nature has arisen only after hundreds of thousands or even millions of years of evolutionary history. We know little of the great many spontaneous emergent species that were not evolutionarily successful and have died off long before they could leave traces of their existence. Since the number of possible maladaptive emergent behaviours greatly exceeds the number of adaptive behaviours, relying on spontaneous co-ordinated behaviour amongst humans to achieve mission outcome may be unwise.

There is, however, a flaw in this argument. Humans are markedly more complex than fish, birds or even wolves. Humans have the ability to think, to self-reflect and to assess their own behaviour in the context of larger objectives (that is, to act in a manner that goes beyond satisfying personal needs).[13] Before acting, humans can discuss issues, create plans, consider the strengths and weaknesses of these plans and can *agree* to co-ordinate their own actions a priori with those of others. A classic example is the formation of spontaneous teams to engage in co-ordinated but unorganized sports (such as street hockey).[14] Nevertheless, it is one thing to say that humans can engage in spontaneous co-ordinated activities, and quite another to say that this spontaneous activity can achieve the levels of co-ordination necessary for the proper arrangement of resources and effort, both in time and in space, to harmonize mission critical effects — that is, to satisfy our definition of co-ordinated action. Arguing that spontaneous co-ordinated action is possible in no way guarantees that it will happen correctly at the appropriate time and in the appropriate context.

Mixing Explicit Control with Emergent Behaviour. We have briefly discussed two general approaches for achieving co-ordinated action: the first imposes external structures and processes for explicitly co-ordinating action, and the second allows co-ordinated action to emerge spontaneously. Both have their strengths and their weaknesses. The first approach may be reliable in known situations but inflexible for situations that are not envisaged; the second approach may adapt to unforeseen circumstances but may do so unreliably and inefficiently. Obviously, some appropriate mixture of both is needed and, indeed, militaries have been using both classes of approach for years. The question therefore becomes: for what situations and under what circumstances should one approach be emphasized over the other?

The question is reminiscent of a tension we have discussed elsewhere: the need to encourage creative command versus the need to control command creativity. Human creativity and will, we have argued, are the two most important characteristics for military command.[15] Creativity is often the only means for solving unforeseen problems, paradoxes and dilemmas (both large and small) during military operations, while will is the source of motivation and commitment necessary to express these creative solutions, especially under arduous operational conditions. But as militaries know, it is inefficient to reinvent the same solutions to problems that have been seen before. Rather, it is better to instantiate these solutions into structures and processes (for example, standard operating procedures, doctrine, defence and weapons systems) and then to use them at the appropriate time. The danger arises when there is an overabundance of pre-defined solutions and an over-willingness to apply them even to problems for which they may be unsuitable — in other words, a tendency for over-control — that eventually translates into organizational rigidity and inflexibility. To achieve co-ordinated action, militaries must balance the requirement to externally control large numbers of resources against the need to allow some of those resources (that is, humans) the freedom to control and to co-ordinate themselves.

We offer the idea of *common intent* as a conceptual framework by which militaries can decide and evaluate the correct mixture of the two classes of approach for achieving co-ordinated action. But before we elaborate on common intent, it is first necessary to discuss intent and to emphasize that the concept of intent intrinsically incorporates both approaches for co-ordinated action. Military organizations, as well as individual commanders, are continually adjusting (often unconsciously) the use of one class of approach versus the other to achieve co-ordinated action through intent.

Intent and Common Intent

Intent is more than an aim or purpose; it is an aim or purpose with all of its associated connotations. Intent conveys the idea of needing to interpret an aim in the context of unforeseen circumstances. For example, a military objective may involve securing and stabilizing an area to allow humanitarian relief efforts. Assuming that the physical resources exist to accomplish this objective, what factors are involved for ensuring that the objective is interpreted correctly by those charged with its completion? Is it simply a matter of giving explicit instructions on how the objective should be accomplished? If so, how explicit should these instructions be? Will it be necessary to elaborate on these explicit instructions? If so, how extensive should the elaborations be? At some point elaboration, amplification and clarification must give way to action. When should this happen? In other words, how much effort must a commander expend to ensure that the connotations of the objective are understood by subordinates? When does commander control give way to subordinate freedom of action? These questions are particularly important when commanders are not familiar with the individual capabilities of their subordinates (for example, during coalition operations) or if the circumstances of the mission are unusual.

The concept of intent includes an explicit portion that contains the stated objective (as well as all of its elaborations) and an implicit portion that remains unexpressed for reasons of expediency but nonetheless is assumed to be understood. This unexpressed implicit intent guides or bounds (but does not direct) the actions of subordinates when faced with unanticipated circumstances. These explicit and implicit aspects of intent roughly correspond to the two general approaches for achieving co-ordinated action. Explicit intent corresponds to the requirement for explicit control, and implicit intent corresponds to the necessity for allowing spontaneous behaviour to emerge consistent with the overall objective. However, for the concept of intent to be useful — that is, for it to contribute to co-ordinated action — it must be shared between one or more individuals. Intent must be *common* between individuals.

In our previous work we had proposed four mechanisms for sharing explicit and implicit intent that were a generalization of Ikujiro Nonaka's four methods for creating knowledge.[16] The two most important of these mechanisms were dialogue for sharing explicit intent and socialization for sharing implicit intent. If commanders shared overt knowledge of the mission objective through dialogue and if they shared tacit knowledge on how to interpret the objective through socialization, then the likelihood of having common intent with their subordinates would be enhanced. We have since realized that dialogue[17] and socialization are necessary, but not sufficient, conditions for creating common intent. There are at least two other important factors: (1) there must be comparable levels of reasoning ability among subordinates for making decisions when neither the time nor the opportunity exists to obtain advice from the commander; and (2) there must be comparable levels of motivation and commitment to achieve mission objectives.

Before we discuss these two additional factors and their importance for common intent, it is first necessary to introduce one last concept that will place much of the discussion thus far into context.

Bounding the Solution Space

In principle, any open-ended problem has an infinity of possible solutions, and the types of problems military members face are almost always open ended. Reducing an infinity of possible solutions to a finite set of practical ones, especially in a short amount of time, is an extremely difficult task. Add to that the difficulty of communicating this reduced set of solutions to other individuals, many of whom may have generated their own set of solutions (which may or may not be consistent with the commander's set), and the potential for mismatch in intent is great. Finally, even if commanders and their subordinates are fortunate enough to work from the same set of solutions, what guarantee is there that the subordinates will translate these solutions into actions that are consistent with each other and that will lead to a co-ordinated effort? How can chaos be avoided? Recognizing that most military solutions are, in fact, (1) quite appropriate to the problem, (2) quite co-ordinated in their effects, and (3) quite successful in their outcome, how does this happen?

The answer is to view the whole issue as one of bounding the (infinite) space of possible solutions. If the number of solutions to a particular mission can be reduced in size by applying a priori principles — principles pertinent to all possible problem types — then for any given problem the solution space is reduced significantly. We would argue that most military forces in democracies have such principles in place and that for the Canadian Forces these a priori principles, at their highest level, are enshrined in the manual *Duty with Honour: The Profession of Arms in Canada.*[18]

Consider Figure 4.1. It depicts an abstract representation of an infinite solution space delineated by three bipolar axes: legal versus illegal solutions, professional versus unprofessional solutions, and ethical versus unethical solutions. If a military organization limits itself only to those solutions that satisfy legal, professional and ethical principles, then the number of acceptable solutions is considerably reduced (Region A in Figure 4.1).[19]

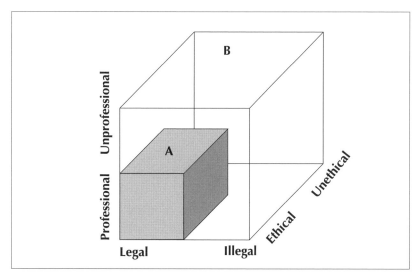

FIGURE 4.1. INFINITE SOLUTION SPACE DIVIDED INTO TWO REGIONS: (A) THE ACCEPTABLE SOLUTION SPACE AND (B) THE UNACCEPTABLE SOLUTION SPACE. NOTICE THAT REGION A IS CONSIDERABLY SMALLER THAN REGION B.

We should note, however, that it is a non-trivial task for organizations to define their principles clearly and consistently enough to avoid placing their members in situations where legal solutions will be viewed as unethical, or professional solutions judged illegal, et cetera. Furthermore, all personnel must understand what these principles are, what they mean, and how they can be used to help solve day-to-day operational problems. Hence, militaries must expend considerable effort and resources towards ensuring that the principles themselves are internally consistent and that they are then taught and reinforced consistently to all members. Typically this process happens over the course of a member's career, especially as he or she gains more authority and assumes more responsibility, and it usually manifests itself as a form of organizational indoctrination.

These a priori principles for bounding the solution space are only three among many possible principles — though these three are the most important and should supersede all others. Within each of the operational environments (for example, army, navy, air force) physical, logistic and safety restrictions may further reduce the number of possible solutions that can be considered.

Commanders must develop their intent within the bounds of a whole hierarchy of guiding principles that limit the types of solutions that they can entertain (see Figure 4.2). If a commander's intent statement implies action that is close to the boundary of one or more guiding principles, then there will be a greater likelihood that subordinates will misinterpret the meaning behind the intent and carry out some action that neither society nor the military organization will tolerate.

We do not mean to imply that the hierarchy of guiding principles exists to control explicitly the actions of commanders. Rather, these principles provide the boundary conditions within which commanders are free to control themselves. This is a critical point, one that may help to further elucidate the differences between the two general classes of approaches for achieving co-ordinated actions discussed earlier. The philosophy behind the first class of approaches is that success is best achieved through explicit instructions for what action should be taken and even, perhaps, for how the action should be undertaken. This assumes that the correct solution to the problem exists and is known ahead of time. Furthermore, it assumes that achieving the objective is simply a matter of

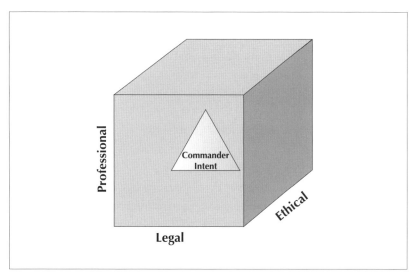

FIGURE 4.2. COMMANDER'S INTENT MUST LIE WITHIN THE ACCEPTABLE SOLUTION SPACE.

implementing the plan of action. Taken to extremes, this philosophy restricts subordinate freedom of action because it discourages the search for alternative solutions.

The philosophy behind the second class of approaches espouses that only guidelines for non-acceptable solutions need to be specified, and that subordinates should search the solution space themselves for acceptable solutions. This approach requires that high agreement exists among participants about what constitutes acceptable versus non-acceptable solutions. If disparity or uncertainty in the solutions set is high among commanders and subordinates, then there will be a greater range of possible actions and, as a result, a smaller likelihood for spontaneous co-ordinated action. We hypothesize that this is one of the major reasons coalition operations are so difficult.

Returning to the concept of intent, we assert that intent allows for both classes of approaches for achieving co-ordinated action. First, it contains an explicitly stated aim or purpose (that is, explicit intent) that constitutes the commander's own predetermination of where the correct solution lies within the acceptable solution space. And second, intent contains that which the commander leaves unsaid but which he or she assumes to be understood by subordinates. This implicit intent carries with it the expectation that subordinates will find their own solutions within (1) the context of the operation as it unfolds and (2) the hierarchy of guiding principles for proper behaviour established by the commander during exercises, by the military profession throughout the subordinates' career, and by the legal and ethical norms of society.

Balancing Explicit and Implicit Intent

It is appropriate to reconsider a question that was asked earlier: When does commander control give way to subordinate freedom of action? This question can now be reformulated in a slightly different form: How much effort should a commander expend in making his intent explicit in order for him to have confidence that his implicit intent is understood by subordinates?

Peder Beausang found that commanders expend considerable effort making intent explicit and that this effort varies over the duration of an operation.[20] He interviewed 12 Canadian and 12 Swedish senior commanders who had experience commanding coalition operations and asked them to draw a graph estimating the amount of effort they expended in making their intent clear during a twelve-month operation. Beausang found that most commanders expended considerable effort making their intent explicit during the first four to six months of an operation, after which their effort decreased significantly towards the end of the operation. Some differences between commanders occurred in the slope of the decrease and whether or not they felt a renewed surge of effort was needed later in the operation, but overall there was considerable consistency. The important point to draw from this research is that commanders felt they could relax their effort to make their intent explicit without feeling that they were compromising the mission. This happened, we hypothesize, because commanders were gaining confidence that the meaning behind their objective(s) was being correctly interpreted by subordinates — that subordinates were correctly inferring the commanders' full intent, both explicit and implicit, when on their own.[21]

Commanders, therefore, must continually estimate the need for explicit control (that is, Class 1 approaches) versus allowing subordinates the freedom of action to solve their own operational problems (that is, Class

2 approaches). Achieving the correct balance is key to achieving co-ordinated action efficiently.

We assert that there are three factors that influence a commander's estimate of the correct balance between explicit and implicit intent: (1) the amount of explicit and tacit knowledge that subordinates share for guiding their actions, (2) the degree of comparability that exists in the reasoning ability of subordinates, and (3) the level of commitment and motivation towards the mission that subordinates share.

Shared Knowledge. To a large extent, we have already discussed this factor. Commanders need to know how well their subordinates understand their explicit intent for the mission, and more importantly, they need to know how well subordinates have internalized the a priori guiding principles that bound proper military behaviour and acceptable military solutions. If commanders are not confident that their explicit intent has been understood adequately — that is, they are not confident that subordinates have understood what to do — then their only recourse is to take the time and explain their intent more fully. But if commanders are not confident that subordinates share even the same guiding principles for acceptable behaviour, then they have a much more daunting task. Not only must they be more explicit about what to do in the mission, but they must be explicit about what not to do, which can be very time consuming since unacceptable solutions greatly outnumber acceptable ones. In other words, if commanders are not confident that their subordinates' solution spaces are sufficiently well bounded, then they will not be confident that spontaneous, acceptable, co-ordinated behaviour will emerge in their absence. Commanders, therefore, must continually assess both the level of overt knowledge about the mission and the level of tacit knowledge about guiding principles that subordinates share for interpreting intent.

As a guide for the type of knowledge commanders should be assessing, we offer David Noble's list of 12 "cognitive enablers" that team members need to be aware of for effective teamwork (see Table 4.1).[22] Though the list may not be complete, it is nonetheless daunting.

Knowledge Enablers for Effective Teamwork	
1	*Goals:* understanding requirements and what a good result looks like
2	*Roles, tasks, and schedule:* includes knowing progress indicators
3	*Relationships and dependencies:* understanding how information and resources impact tasks and how tasks impact goals
4	*Team members' backgrounds and capabilities:* understanding what others can and will do under various circumstances. Key to trust.
5	*Team "business rules":* knowing the agreed-upon procedures for information, helping each other, and resolving conflict
6	*Task knowledge:* knowing how to do your job
7	*Activity awareness:* knowing what others are doing, how busy they are, and whether they are working on the right thing
8	*The external situation:* understanding adversaries and competitors and how they can impact team success
9	*Task assessment:* tracking task progress
10	*Mutual understanding:* knowing the extent to which team members agree or disagree
11	*Plan assessment:* understanding whether the team's plan will still work
12	*Decision drivers:* understanding decision drivers, deadlines, handling uncertainty, and who to consult

TABLE 4.1. KNOWLEDGE ENABLERS FOR EFFECTIVE TEAMWORK

SOURCE: D.F. NOBLE, "UNDERSTANDING AND APPLYING THE COGNITIVE FOUNDATIONS OF EFFECTIVE TEAMWORK" (VIENNA, VA: EVIDENCE-BASED RESEARCH INC., 2004)

Comparable Reasoning Ability. Shared knowledge of the commander's objective, as well as shared knowledge of the acceptable solution space, is not sufficient for co-ordinated action. Shared knowledge establishes the initial conditions for decision making and co-ordinated action, but it does not address subordinates' ability to make decisions or to initiate co-ordinated action itself.

Recent theories of group interactions and team decision making have placed much emphasis on the concepts of *shared mental models*[23] and *shared situational awareness.*[24] These theories assert that if team members develop models of each other and of each others' tasks, and if they share knowledge of the circumstances of the mission as well as the mission objective, then team members will be able to anticipate each other's needs, predict future problems and co-ordinate their actions. The original work on shared mental models grew out of the Tactical Decision Making Under Stress (TADMUS) program, which studied decision making in the operations room onboard U.S. Navy warships; the work on shared situational awareness drew from studies of aircrew on flying missions. Each research effort developed its theoretical perspective from studying relatively small teams with homogeneous members of roughly comparable reasoning ability. Over the past decade, the shared mental models theory and the shared situational awareness theory have been very successful, amassing much corroborating evidence and analytical support. These theories support the fundamental imperative of sharing knowledge among team members; however, they assume that team members share comparable levels of reasoning ability and that team members are roughly equivalent in their ability to make inferences, draw conclusions and take appropriate action. As a result, these theories are inadequate for situations where teams have diverse composition and capability and face uncertain operational situations that go beyond their shared mental models.

As most commanders know, understanding and interpreting intent is not a simple exercise. It requires significant cognitive skills such as sustained attention and the storing and accessing of knowledge from short- and long-term memory; it also requires inductive and deductive inferencing, forming comparisons, and drawing conclusions. In short, it requires sophisticated reasoning ability. Given that modern coalition operations consist of teams from different rank structures (for example, officers, non-commissioned members, civilians), different services (for example, army, navy, air force) and different nationalities (for example, coalition operations), to what extent do comparable reasoning abilities exist among such diverse teams of heterogeneous members? Furthermore, to what extent is the coalition commander aware of these differences and what must he or she do to accommodate them?

We have argued that achieving co-ordinated action requires the correct balance of explicit control and spontaneous emergent behaviour. We have also argued that the former is an aspect of explicit intent and that the latter depends on the establishment of guiding principles for bounding the acceptable solution space, which then allows commander intent to be interpreted correctly. But it is one thing to know what must be done (that is, understand explicit intent) under circumstances that may have been envisaged by the commander, and it is quite another to be able to solve a new operational problem in the commander's absence. Even if a subordinate has internalized the correct guiding principles and, as a result, is inside the space of acceptable solutions, that solution space nonetheless remains vast and must be searched (quickly), using relatively sophisticated reasoning abilities. Not every member of large heterogeneous teams will have that kind of intellectual capability. Furthermore, it may be *unreasonable* to expect that every member of a military organization achieve this level of reasoning ability.

Faced with this reality, commanders are left with three strategies. First, commanders should identify, as soon as possible, those individuals who demonstrate a competence for thinking a problem through. These individuals should occupy key roles in the commander's team. Second, commanders should match the difficulty of the task to the intellectual ability of the member. Not all problems are equally onerous, nor are they all equally critical for mission success. Commanders should be judicious in the formation of teams and the tasks those teams are assigned. Third, commanders should ensure that subordinate commanders engage in similar kinds of strategies — that is, carefully choose their teams and allocate tasks according to competence. Of course, these strategies are not new; indeed, in our informal discussions with senior commanders, most have acknowledged using precisely these approaches. What we have provided, we hope, is a theoretical framework that situates these strategies within the larger objective of establishing common intent to achieve co-ordinated action.

Shared Commitment and Motivation. The third factor that influences the amount of effort a commander will expend to make his intent explicit involves the overall level of motivation and commitment that subordinates display towards achieving the mission. We had mentioned earlier that will is a key component of command. We define *will* as "diligent purposefulness," and for achieving common intent it is as important as shared knowledge and comparable reasoning ability.

Interpreting intent requires effort. So does co-ordinated action. It is almost inconceivable that a military operation could be successful without considerable motivation and commitment being displayed by those who carry it out. Indeed, considering the range of human tragedy that militaries are exposed to, and the inevitable dangers their members must face, it is a wonder that members exhibit as much initiative and commitment as they do. For example, it is quite conceivable that subordinates could feign ignorance on some aspect of the commander's objective and decide to carry out the letter rather than the spirit of the intent. A common example of this behaviour in the civilian sector is the invocation of work-to-rule job actions. Here the collective will of the work force is channelled into restricting effort, thus seriously affecting the smooth functioning of the organization. Thankfully, such collective obstructionism is rare among militaries (except, perhaps, as a prelude to mutiny), but, as with innate reasoning ability, there may nonetheless be significant individual differences in motivation and commitment among members. Commanders, therefore, must pay close attention to morale and esprit de corps among their teams.

Extolling the virtues of unit morale and commitment may seem a trifle obvious, especially to a military reader, but emphasizing its importance within the context of explicit and implicit intent and the two classes of approaches for achieving co-ordinated action is novel and constructive. It stresses the effortful nature of interpreting intent, and it underlines the necessity of *fuelling* spontaneous co-ordinated action. Like all emergent behaviour, co-ordinated action needs a source of energy. Commanders, therefore, must tap the creative will of their subordinates at the same time as they explicate their intent. They must realize that conveying intent is more than conveying a concept of operation; it is also an opportunity to energize subordinates, to motivate independent thought and action, and to make members commit to working together. As in all aspects of military life, the importance of the leadership skills of the commander cannot be overemphasized.

Common Intent

Earlier we had defined C^2 as the establishment of common intent to achieve co-ordinated action. We then discussed the nature of co-ordinated action and delved into the concept of intent itself, including the three factors to be considered in balancing the need for explicit versus implicit intent. We now return to the idea of common intent.

As an independent human with a creative will, each military member has the capacity to develop his or her own intent. Although we have spoken in this chapter as if only commanders could have intent, the fact is that all humans have goals, plans and objectives. Even junior military members with little or no experience can reason about situations, can have opinions, can ask questions and can seek information about the mission. The task of the commander is to harness this human potential and to focus it with his or her own intent. Common intent is achieved when there is a single shared objective, together with a clear understanding of how that objective can be attained. As such, common intent is an idealized concept where maximum overlap, with minimum dispersion, exists between the intent of a commander and the intents of his subordinates (see Figure 4.3).

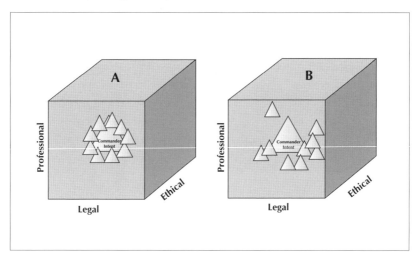

FIGURE 4.3. COMMANDER AND SUBORDINATE INTENT (A) TIGHTLY CLUSTERED OR (B) POORLY CLUSTERED WITHIN THE ACCEPTABLE SOLUTION SPACE.

In reality, complete common intent is probably impossible. It would require every member of a unit to share the same knowledge as the commander, to incorporate the same guiding principles, to possess the same reasoning ability and to experience similar levels of commitment. In essence, it would require a unit consisting entirely of clones of the commander. This is, of course, unrealistic.[25]

Recall that co-ordinated action can be achieved using explicit control (Class 1 approaches) or spontaneous emergent behaviour (Class 2 approaches), and that some combination of the two is almost always necessary. Complete common intent, as described above, may specify the best conditions for the emergence of spontaneous co-ordinated action, but, to the extent that commanders (inevitably) are unable to achieve this ideal, they must compensate with Class 1 approaches. As we saw with Beausang's study, spending more time explicating intent with subordinates is a principal means of compensating for heterogeneous knowledge, reasoning ability and commitment. Other techniques include hand-picking team members and matching task complexity to team competency. But there is also a host of other control mechanisms commanders can use, mechanisms that are embedded in the fabric of the military organization. For instance, the chain of command and rank structure can be invoked to greater or lesser degrees depending on commander's need to be directive.

Figure 4.4 summarizes the relationship between Class 1 and Class 2 approaches, shared explicit and implicit intent, common intent and organizational structure. The extreme left and right of the figure represent predominantly Class 1 and Class 2 approaches, while the centre area reflects a varying mixture of the two. As we progress from the left side of the figure, where shared explicit intent dominates over shared implicit intent, to the right of the figure where the reverse is true, common intent becomes greater due to the greater overlap between commander and subordinates on the three intent factors (knowledge, reasoning ability and commitment). We hypothesize that there is a central segment of this figure that represents an optimal area within which C^2 organizations should reside. We should add, however, that there are also other factors that will influence where on Figure 4.4 a military will lie. For example, a military's willingness to tolerate risk will impact how decentralized its C^2 structure will be, or a society's willingness to tolerate casualties may influence the level of autonomy that commanders are given to make choices.

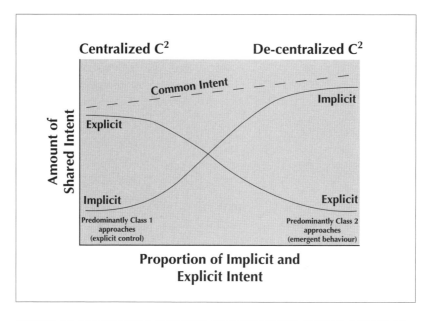

FIGURE 4.4. RELATIONSHIP BETWEEN COMMON INTENT, SHARED INTENT AND ORGANIZATIONAL STRUCTURE

Contemporary command philosophies like *network-enabled operations* (NEOps) or *network-centric warfare* (NCW)[26] assume that if a large number of humans are linked together as nodes in a network, and if common and accurate information about the operation is made available to them, self-organizing behaviour will emerge that will yield unsurpassed knowledge superiority and speed of response. What the NEOps philosophy fails to recognize, however, is that detailed and accurate information is only one necessary condition for (self-) synchronized, co-ordinated action. NEOps assumes that guiding principles for defining acceptable solution spaces are known and are embodied by military members; it assumes that comparable abilities for analyzing a situation and making decisions exist; and it assumes that all members are as committed to achieving the objective as they should be. These are unwarranted assumptions given the fact that theorists like Frank Fukkuyama view a network as "a moral relationship of trust…[among] a group of individual agents who share informal norms and values,"[27] rather than as a simply physically interconnected node for exchanging information.

Conclusion

Responsible command and control for achieving military mission objectives is among the most complex, difficult, stressful and dangerous challenges facing any commander. To suggest that a single theoretical framework can capture the full breadth and depth of this challenge is perhaps naïve; however, theories, even incomplete ones, are invaluable for adding conceptual clarity to a topic.

Our definition of C^2 emphasizes the critical importance of establishing common intent among military members — a common intent that we believe is necessary for achieving co-ordinated action. We assert that an important aspect of the military operational art is to balance Class 1 versus Class 2 approaches for achieving co-ordinated action. A mission that relies too heavily on the first approach may suffer from rigid over-control; a mission that relies too much on the second may spiral into chaotic inefficiency. The philosophy of mission command suggests that militaries, in general, should adopt a C^2 structure that lies as far right on the continuum in Figure 4.4 as possible, without sacrificing efficiency. In reality, the actual blend of Class 1 and Class 2 approaches will depend upon a number of military and extra-military factors such as political sensitivity, threat to home nation, size of the operation, logistics, operational tempo, and interaction with non-governmental organizations. Regardless of where on the C^2 continuum (that is, Figure 4.4) a military mission lies, individual commanders can nonetheless maximize common intent within their limited span of influence. By paying attention to the amount of explicit and tacit knowledge subordinates share, by assessing their ability to reason based on that knowledge, and by influencing their overall level of motivation and commitment to achieve the objective, a commander can take full advantage of the potential for common intent that resides in his or her subordinates.

Table 4.2 is a cursory example of how a commander can use the three factors to diagnose the potential among his subordinates for achieving common intent. Each factor is identified as either maximally present or minimally present, and all eight combinations across the three factors are listed. We hypothesize that the greatest potential for achieving common intent exists when all three factors are assessed by the commander as being maximally present. Conversely, when all three factors are minimally present, the commander will have significant challenges achieving any kind of action (co-ordinated or otherwise). The six combinations in between these two extremes offer varying levels of common intent potential. In general, minimum motivation and commitment implies leadership challenges. Minimally shared explicit and tacit knowledge implies that subordinates may fail to understand mission objectives as well as fail to operate within the acceptable solution space. A minimum amount of subordinate reasoning ability implies that their ability to draw inferences in the absence of the commander will be hampered. Each combination of factors requires different responses from the commander in order to yield the greatest likelihood of achieving co-ordinated action.

Shared Knowledge	Comparable Reasoning Ability	Shared Commitment and Motivation	Impact on C^2
Maximum	Maximum	Maximum	Greatest potential for establishing common intent
Maximum	Maximum	Minimum	Wasted potential for common intent (leadership issue?)
Maximum	Minimum	Maximum	Some potential for common intent; will need to rely on very detailed plans and explanations
Maximum	Minimum	Minimum	Poor potential for common intent; leadership and detailed plans required
Minimum	Maximum	Maximum	Good potential for common intent if guiding principles for appropriate action exist (means more effort needed for explicating objective); if shared guiding principles do not exist, unacceptable solutions are a possibility
Minimum	Maximum	Minimum	Little potential for common intent; leadership and very detailed, explicit intent are required
Minimum	Minimum	Maximum	Dangerous common intent; over zealousness may lead to unco-ordinated chaos with high potential for unacceptable solutions
Minimum	Minimum	Minimum	Least potential for establishing common intent

TABLE 4.2. A METHOD FOR DIAGNOSING THE POTENTIAL FOR ACHIEVING COMMON INTENT AMONG SUBORDINATES

From a military perspective, attempting to dissect C^2 may seem overly analytical and sterile. After all, military commanders have been "doing" C^2 more or less successfully for hundreds of years. Some may argue that too much analysis, especially if it is incomplete, may actually get in the way of excelling in military command and other command-related activities like the operational art. From a scientific perspective, some researchers may view the theoretical framework we propose as too loose and imprecise. For example, we have only hinted at the mechanisms underlying the two approaches for co-ordinated action, and we have only asserted, without empirical justification, that three factors are necessary for common intent. Both criticisms have merit, yet both criticisms suffer the same shortcoming. Both assume that only complete knowledge can further the practice and understanding of a field or discipline. We offer this theoretical framework as a means of spurring debate among military practitioners and of encouraging rigorous experimentation among researchers.

NOTES

1   Society's expectations for the Canadian Forces are rooted in the values of the Canadian people as mentioned in Canada, Department of National Defence (DND), *Duty with Honour: The Profession of Arms in Canada* (Kingston, ON: Canadian Defence Academy, 2003).

2   See for example DND, *Canadian Forces Operations*, B-GJ-005-300/FP-000 (2004).

3   C. McCann and R.A. Pigeau, "Clarifying the Concepts of Control and of Command," in Proceedings of the Command and Control Research and Technology Symposium, Newport, RI (29 June–1 July 1999).

4   R.A. Pigeau and C. McCann, "Re-defining Command and Control," in *The Human in Command: Exploring the Modern Military Experience*, eds. C. McCann and R.A. Pigeau (New York: Kluwer Academic/Plenum Publishers, 2000), 163–84.

5   Ross Pigeau and Carol McCann, "Re-conceptualizing Command and Control," *Canadian Military Journal* 3, no. 1 (Spring 2002), 53–63.

6   DND, *Canadian Forces Operational Planning Process*, B-GJ-005-500/FP-000 (6 November 2002), http://www.dcds.forces.gc.ca/jointDoc/pages/j7doc_docdetails_e.asp?docid=16

7   R.A. Pigeau and C. McCann, "Putting 'Command' back into Command and Control: The Human Perspective," paper presented at the Command and Control Conference, Congress Centre, Ottawa (25 September 1995).

8   C. McCann and R.A. Pigeau, "Taking Command of C^2," in Proceedings of the Second International Command and Control Research and Technology Symposium, Market Bosworth, Warwickshire, UK (23–25 September 1996).

9   See E. Jantsch, *The Self-Organizing Universe* (Toronto: Pergamon Press, 1980); and S. Kaufmann, *At Home in the Universe: The Search for the Laws of Self-Organization and Complexity* (New York: Oxford University Press, 1995).

10   G. Nicolis and I. Prigogine, *Self-Organization in Non-Equilibrium Systems* (New York: J. Wiley & Sons, 1977).

11   P. Grassé, "La reconstruction du nid et les coordinations inter-individuelles chez Bellicositermes Natalensis et Cubitermes sp. La theorie de la Stigmergie: Essai d'interpretation du comportment des termites constructeurs," *Insectes Sociaux* 6 (1959), 4181.

12   It is very difficult to discuss insect behaviour without using human concepts like *social* and *knowledge*, though little of what humans mean by the words *social* and *knowledge* apply to insects.

13   J. Q. Wilson, *The Moral Sense* (New York: Free Press Paperbacks, 1993).

14   Very little is known about *how* spontaneous co-operative behaviour emerges among humans, although complexity theorists R. Axelrod and M.D. Cohen, *Harnessing Complexity: Organizational Implications of a Scientific Frontier* (New York: Free Press, 1999), are attempting to outline the general conditions and mechanisms.

15   Pigeau and McCann, "Re-conceptualizing Command and Control," 53–63.

16   Pigeau and McCann, "Re-defining Command and Control," 163–84.

17   Dialogue in this case means more than just verbal dialogue and includes written text, figures and non-verbal expressions.

18   DND, *Duty with Honour*.

19   For ease of depiction, the acceptable solution space in Figure 4.1 is approximately one-eighth the size of the entire solution space. In reality, the acceptable solution space is probably considerably smaller since the number of illegal, unprofessional and unethical solutions likely corresponds to more than half of each axis.

20   P. Beausang, "The Role of Intent and the Ideal Command Concept in Military Command and Control: Canadian and Swedish Commanders' Perspectives," Stockholm, FOI — Swedish Defence Research Agency, report # FOI-R-1069-SE (2004).

21   The reasons could be that commanders were learning about the strength and limitations of their subordinates, and conversely, subordinates were learning the boundary conditions for acceptable solutions from their commanders.

22   D.F. Noble, "Understanding and Applying the Cognitive Foundations of Effective Teamwork" (Vienna, VA: Evidence-Based Research, 2004), http://www.stormingmedia.us/08/0833/A083324.html.

23   E. Salas, C.A. Bowers et al., "Military Team Research: 10 Years of Progress," *Military Psychology* 7, no. 2 (1995), 55–75.

24   M.R. Endsley, B. Bolte et al., *Designing for Situation Awareness: An Approach to Human-Centered Design* (London: Taylor & Francis, 2003).

25   Cloning is not only unrealistic, it is undesirable. We should stress that though complete common intent may ensure co-ordinated action, it would not guarantee that the action taken is necessarily correct for mission success. Indeed, cloning the commander, even if this were possible, would likely also guarantee the emergence of "group think" behaviour. *Groupthink* is a phenomenon (see I. Janis, *Victims of Groupthink: A Psychological Study of Foreign-Policy Decisions and Fiascoes* [Boston: Houghton Mifflin, 1972]) whereby individual initiative for finding alternative solutions is (unconsciously) discouraged for the sake of achieving group consensus. J. Surowiecki in his book *The Wisdom of Crowds: Why the Many Are Smarter than the Few and how Collective Wisdom Shapes Business, Economics, Societies, and Nations* (New York: Doubleday, 2004) reviews the scientific literature that argues for diversity in groups — diversity in background, intelligence, skills et cetera — because such diversity stimulates the generation of "random" ideas and alternative solutions to problems. Tension exists, therefore, between maximizing the conditions for achieving co-ordinated action and maximizing the conditions for finding the best solutions. Part of the commander's operational art is also to discover the correct balance between the two objectives.

26   *Network- Enabled Operations* is Canada's version of the US's *Network Centric Warfare* concept. See Allan English et al. "Beware of Putting the Cart Before the Horse: Network Enabled Operations as a Canadian Approach to Transformation," DRDC Toronto, Contract Report CR 2005-212 , (19 July 2005) for a discussion of networked operations.

27   Frank Fukuyama, *The Great Disruption: Human Nature and the Reconstitution of Social Order* (New York: Free Press, 1999), 199.

LEADERSHIP AND COMMAND

CHAPTER 5

CHAPTER 5

WE FIGHT AS ONE?
THE FUTURE OF THE CANADIAN FORCES' JOINT OPERATIONS GROUP

Colonel C.J. Weicker

The side that has superiors and subordinates united in purpose will take the victory.

Sun Tzu[1]

The Canadian Forces (CF) command and control structure has evolved over the last decade to meet the demands of domestic and international operations, as well as force reductions caused by the 1994 Defence White Paper.[2] During the 1990 Oka Crisis and the 1991 Persian Gulf War, National Defence Headquarters (NDHQ) developed a dedicated joint staff to command operations at the strategic level.[3] At the operational level, the ad-hoc creation of a joint headquarters in Bahrain prompted the CF to consider forming a permanent capability for future operations.[4]

In 1994, Armed Forces Council directed the development of a CF operational-level command and control capability.[5] In 1996, NDHQ posted a 35-person, joint headquarters cadre staff to the Army's 1st Canadian Division Headquarters, which was made responsible for providing a deployable joint headquarters.[6] Several domestic operations tested the joint headquarters' capability.[7] During the same period the CF created a project to evolve the 1st Canadian Division Headquarters into the CF Joint Operations Group (JOG), whose role would be to "provide operational-level command and control capabilities for the CF." Since its formation in June 2000, the JOG progressively developed its capabilities and reached full operational capability three years later. The motto it adopted, "We fight as one," was to signify its joint war-fighting role.[8]

Since the paper on which this chapter is based was written in 2003, the JOG has been disbanded and its members posted into the new commands created as part of CF transformation activities started in 2005.

THE OPERATIONAL ART

109

Nevertheless, the strategic- and operational-level command and control issues raised in this chapter are still relevant as the command and control arrangements for the new CF commands created as a result of transformation initiatives started in 2005 are based on lessons learned and procedures devised in the ten or so years in which the JOG was created and existed. Therefore, the JOG remains a valuable case study in the creation and evolution of Canadian command and control arrangements.

During the period covered by this chapter, within NDHQ, the Deputy Chief of the Defence Staff (DCDS) was responsible to plan and control operations. During that period, due to the lack of dedicated staff and the high tempo of operations, many of the strategic-level functions were not done well, or done at all, in order for the essential operational-level functions to be completed. Therefore, when the JOG became formally operationally capable, the CF should have adjusted the responsibilities, processes and structure of NDHQ and the JOG, with respect to CF contingency operations, to make full use of the JOG's capabilities.

This chapter argues that once the JOG became operationally capable it should have been given full authority and responsibility for the command and control of all CF contingency operations. The first part will explain some key terms and command and control principles before examining the CF's concepts and doctrine for the command and control of contingency operations. The second part will then examine the reality experienced by the JOG on three recent contingency operations. Based on this analysis, potential options to resolve the issues identified here will be evaluated in the third part. It will be shown that the best option at the time was to transition the JOG headquarters to become the sole CF operational-level headquarters responsible for contingency operations.

Theory

> Theory exists so that one does not have to start afresh every time sorting out the raw material and ploughing through it, but will find it ready to hand and in good order. It is meant to educate the mind of the future commander, or, more accurately, to guide him in his self-education; not accompany him to the battlefield.[9]
>
>   Carl von Clausewitz

A review of the CF theory of command and control needs to first start with some understanding of the CF principles of command and control. The March 2003 draft of the CF Doctrine manual lists six principles of command for the CF (Table 5.1). These principles are based on a CF philosophy of command where commanders must ensure that their subordinates understand their intentions and the assigned mission. In turn, the subordinates will be given "maximum freedom of action," once orders are given, and sufficient resources to decide how to best achieve their mission.

A review of the Environmental (that is, army, navy and air force) doctrine shows that only the Canadian Army has expanded on these command principles. Its Command manual articulates five fundamentals of command based on the need to develop trust and mutual understanding between commanders and subordinates at all levels. A comparison of the CF command principles and Army command fundamentals is shown in Table 5.1.

CF Principles of Command	Army Fundamentals of Command
Unity of Command. A single, clearly identified commander will be appointed for each operation. He or she has the authority to direct and control the committed resources and is responsible and accountable for success or failure.	**Unity of Effort.** Commanders must impart a clear sense of purpose to their subordinates. Subordinates must understand the intent of their immediate superiors and those two levels up. This unity of purposes at three levels of command promotes mutual understanding and allows subordinates to act purposefully in an unexpected situation.
Delegation of Authority. Commanders may delegate all or part of their authority if the scope and complexity of an operation requires it. How much authority is delegated, and to whom, must be clear.	**Decentralized Authority.** Decentralizing decision making includes setting decision thresholds as low as possible to allow for rapid decision making and reduced flow of information up the chain of command. This requires delegation of specific authorities. A commander who delegates authority for action to a subordinate is required to furnish that subordinate with sufficient resources.

...continued on next page

CF Principles of Command	Army Fundamentals of Command
Freedom of Action. Once the mission is established and orders given, maximum freedom of action is given to subordinate commanders.	**Timely and Effective Decision Making.** Commanders must be capable of operating efficiently in an environment of great uncertainty. Commanders must be able to make sound and timely decisions faster than an adversary.
Chain of Command. The command structure is hierarchical and must be clear and unequivocal. Bypassing levels of command in either direction is only justified in exceptional circumstances.	**Trust.** A superior needs to not only earn the trust of his subordinates but also place his trust in them. The basis of this two-way trust is shared implicit intent, which enhances mutual understanding.
Continuity of Command. A clear succession of command, well understood at all levels, is required.	**Mutual Understanding.** Commanders understand the issues and concerns facing their subordinates based on shared perception of military doctrine.
Span of Control. The assigned resources and activities must be such that one person can exercise effective command and control.	

TABLE 5.1. COMPARISON OF CF PRINCIPLES OF COMMAND AND ARMY FUNDAMENTALS OF COMMAND

SOURCES: DND, *CANADIAN FORCES DOCTRINE* (THIRD DRAFT), B-GJ-005-000/AF-000 (MARCH 2003); AND DND, *COMMAND*, B-GL-300-003/FP-000 (21 JULY 1996)

For the purposes of this chapter, the principles and fundamentals have been amalgamated into one list at Table 5.2. This new list, though not officially part of CF doctrine, will be used later to evaluate command and control aspects during contingency operations.

Amalgamated CF Command Principles and Fundamentals
Unity of Command A single, clearly identified commander will be appointed for each mission. The commander has the authority to direct and control the committed resources. The commander is responsible and accountable for the success or failure of the mission.
Chain of Command The chain of command and the related authorities must be clear and tailored for the mission. A clear succession of command, well understood at all levels, will be implemented to ensure proper command authority throughout the conduct of the mission.
Decentralized Authority Decentralizing decision making includes setting decision thresholds as low as possible to allow for rapid decision making and reduced flow of information up the chain of command. This requires delegation of specific authorities. A commander who delegates authority for action to a subordinate is required to furnish that subordinate with sufficient resources. Once the mission is established and orders given, maximum freedom of action is given to subordinate commanders.
Span of Control The assigned resources and activities must be such that a single commander can exercise effective command and control, even in crisis and war.
Unity of Purpose Commanders must impart a clear sense of purpose to their subordinates. Subordinates must understand the intent of their immediate commander and of the superior commanders two levels up the chain of command. This unity of purposes at three levels of command promotes mutual understanding and allows subordinates to act purposefully in an unexpected situation without reference to the commander.
Mutual Understanding and Trust Commanders understand the issues and concerns facing their subordinates based on shared perception of military doctrine, which enhances mutual understanding. A superior needs to not only earn the trust of his subordinates but also place his trust in them. The basis of this two-way trust is shared implicit intent.

TABLE 5.2. AMALGAMATED CF COMMAND PRINCIPLES AND FUNDAMENTALS

In addition to these principles, it is important to distinguish between the two types of CF operations. Routine operations are those for which one of the CF Environments has been specifically tasked, organized and equipped; these operations are normally commanded from one of eight operational-level headquarters[10] across Canada under command of one of the Environments. Contingency operations are the remaining CF operations and can be conducted either domestically or internationally;[11] they are commanded by the Chief of Defence Staff (CDS) with assistance from the DCDS. Contingency operations are normally joint and combined in nature and, therefore, require detailed planning and close control to ensure mission success. The JOG has been designated as the deployable operational-level headquarters for contingency operations. With those principles and definitions in mind, let us turn to the current CF concepts and doctrine related to the strategic and operational levels of command and control.

The source for CF concepts is the 1994 Defence White Paper.[12] *Strategy 2020*, published in 1999, builds on the White Paper and provides the overall strategic direction for the CF. It describes certain objectives to meet this strategy, one of which is related to command and control: *Globally Deployable*. The five-year target for this objective is to "complete the conversion of the Joint Force Headquarters to a deployable C⁴I (command, control, communications, computers and intelligence) organization capable of national command and logistic support at the operational level of war."[13] This target was achieved with the JOG in June 2003; however, the development of the support command and control concepts did not progress at the same rate.

In 1993, the Vice Chief of Defence Staff (VCDS), who is responsible for CF strategic concepts, was developing a new taxonomy for concept development as shown at Figure 5.1.

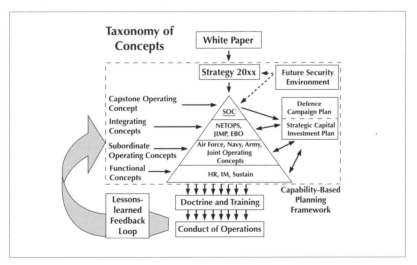

FIGURE 5.1. TAXONOMY OF CONCEPTS FROM A PRESENTATION BY LCOL WAYNE EYRE, DDA 3, TO THE SOC DEVELOPMENT RETREAT, 24–26 SEPTEMBER 2003

The new capstone document, *Strategic Operating Concept*, is currently being written. It will guide the development of other operational concepts, including subordinate (environmental and joint), integrating[14] and functional concepts.[15] In terms of joint command and control, the draft document states: "the CF must adopt, at the strategic and operational levels, a more agile Joint Command and Control.... To implement [it] the CF must review roles and responsibilities currently carried out by the various headquarters at the strategic and operational levels."[16] In addition, the draft subordinate CF Joint Operating Concept 2012, being written by the NDHQ J7 staff, states that by 2012 the Joint Operations Group will be solely responsible for the conduct of operations, whether routine or contingency, in the future.[17] There is a clear indication that the CF intends to eventually make the JOG more responsible for the conduct of CF operations. What is not clear is what changes will be required and how they will be implemented. To assist in answering those questions, it is important to examine the CF's command and control doctrine.

In April 2003, the CF Doctrine Board approved a new CF doctrine hierarchy, based on the continental staff system,[18] with a draft of the CF Doctrine manual, *Canadian Forces Doctrine*, as the capstone document. As of October 2005, this manual was still listed as the capstone document.[19] The draft manual contains the current CF practices in the areas of command and control at both the strategic and operational levels. An examination of the manual's areas of command and control responsibilities, planning and execution will help formulate options for the future of the JOG.

The first area to examine is the split of responsibilities for command and control at the strategic and operational levels. To begin with, it is important to understand the definitions of these two levels and their fundamental differences. The *strategic level* of war is defined by NATO as "the level of war at which a nation or group of nations determines national or multinational security objectives and deploys national, including military, resources to achieve them."[20] The *operational level* is defined as "the level of war at which campaigns and major operations are planned, conducted and sustained to accomplish strategic objectives within theatres or areas of operations."[21] The NATO capstone manual, *AJP-1: Allied Joint Doctrine*, further clarifies the levels of war by stating:

The key to delineation is that normally strategic authority allocates objectives and resources, setting necessary limitations; while, at the operational level, the commander orders the activities of his assigned formations in pursuit of his own plan of campaign. At the tactical level, commanders employ units for combat in order to achieve the military objectives of the campaign.[22]

The CF Doctrine manual states:

The military strategic level is concerned with determining the military strategic goals and the desired end-state, by crafting strategy, allocating resources and applying constraints as directed by the political leadership. The operational level links the strategic and tactical levels. The focus at this level is on operational art; it is at operational level that major operations are planned, conducted and sustained to accomplish strategic goals.[23]

Thus in accordance with NATO and CF doctrine, NDHQ, at the strategic level, is responsible for crafting the strategy, while the JOG, at the operational level, is responsible for developing and implementing the campaign plan of an operation. But, as we will see next, that clear distinction for planning is not used in the CF Doctrine manual.

The CF Doctrine manual, in fact, introduces a new concept of strategic campaign planning. It states that "[c]ampaign planning is concerned with defining the strategic conditions which determine success, translating policy goals into military strategic objectives, assigning operational level command, imposing limitations and allocating resources. Campaign planning at NDHQ is confined, as far as is practicable, to the strategic level, leaving operational level activities to the designated [Task Force Commander]."[24] This statement also appears in the CF Operations manual, but it refers to the development of strategic directives and not campaign plans.[25] The Operations manual contains an annex on *strategic* campaign planning, which is taken almost verbatim from NATO documents for the operational art and campaigning, except for the inclusion of strategic in the title. It would appear that this is an attempt to amalgamate operational-level campaign planning with strategic mission analysis and direction within NDHQ. Since there is no explanation of operational campaign planning, CF doctrine appears to suggest that this would not be one of the above activities performed by the operational headquarters, like the JOG. Clearly this variance from NATO doctrine needs to be reviewed and explained in the next draft of the manual.

A third area to examine is the command and control of a task force during contingency operations. When a contingency operation is authorized, the Environmental Chiefs of Staff will be tasked to provide forces. Once they declare the forces operationally ready, the forces will be transferred under operational command of the CDS. The CDS will then, at an appropriate time, transfer the forces under operational command or control of the Task Force Commander.

At the operational level, a Task Force Headquarters, normally provided by the JOG, is deployed to support the Task Force Commander. If forces are provided by more than one Environment, the Task Force will be designated a Joint Task Force (JTF), and the commander called a Joint Task Force Commander (JTFC). The CF Doctrine manual makes passing reference to the fact that the JOG also provides a logistic support capability through the CF Joint Support Group and communications support to all deployed missions through the CF Joint Signal Regiment. There is no information on the role of the JOG in areas such as strategic reconnaissance, liaison, mission activation and mission closeout.[26]

The NDHQ joint staff provides strategic-level oversight and administrative control of CF operations. The strategic intelligence (J2), operations (J3), plans (J5), training and doctrine (J7), and civil-military affairs (J9) staffs are found within the DCDS Group. The remaining support staff functions, such as personnel (J1), logistics (J4), communication and information systems (J6) and finance (J8), and specialists are provided by other organizations within NDHQ. The Joint Staff Action Team, headed by the Chief of Staff J3 (COS J3), co-ordinates the solution of problems associated with CF operations. It is important to note that since there is no equivalent Chief of Staff J4 to integrate the support functions, the COS J3 is solely responsible for integrating all of the staff functions.

In terms of actual control of operations, the CF Doctrine manual states, "Canada's command and control structure for operations differs from those of our major allies in one important respect: most CF operations are controlled directly from the national headquarters rather than from a subordinate, single-service or joint headquarters established for that purpose."[27] No explanation is provided for this major variance between the CF and its allies.

In summary, the new CF doctrine, which is based on actual practice, varies from NATO doctrine in that it merges the strategic- and operational-level planning and control functions at NDHQ. Though regular co-ordination occurs, the NDHQ joint staff lacks unity of command since the majority of the joint staff are not in the formal chain of command of the DCDS. The absence of a chief of staff for support functions puts unnecessary burden on the COS J3 to integrate both operations and support aspects in planning and controlling operations. Therefore, CF command and control doctrine is inconsistent with the emerging concept of transferring to the JOG the responsibility for command and control of both routine and contingency operations by no later than 2012. An examination of recent operations will demonstrate that the need to conduct this transfer of responsibilities from NDHQ to the JOG is urgently required.

Practice

> By analyzing historical command [and control] systems at work we may hope to gain a better idea of how it was done, successfully or otherwise.
>
> Martin van Creveld[28]

The Joint Operations Group played an important role in the command and control during recent CF contingency operations. Though it was formed in 2000, its capabilities were built on the practical experiences of the 1st Canadian Division Headquarters, stretching as far back as December 1991 when the division headquarters mounted and deployed as the Canadian Joint Force Somalia headquarters. However, there were many improvements after that time, and thus this examination will focus on three recent operations. Before identifying these operations, it would be useful to first describe the actual role and capabilities of the JOG.

The Joint Operations Group was created by the Defence Services Program's Project Number 2001, *CF Joint Headquarters / Joint Task Force Headquarters*. It was a formation that comprised two units: the Joint Headquarters and the Joint Signal Regiment. The Joint Headquarters was a permanent, operational-level joint staff with representation from all the staff branches. The Joint Signal Regiment provided dedicated intelligence, communication and information systems, and combat service support capabilities to the Joint Headquarters.[29]

The JOG was a deployable operational-level command and control capability, which was capable of performing various roles during CF contingency operations from crisis up to and including war fighting. The JOG was designed to form a JTF headquarters that would operate in one of two possible roles depending on whether operational command was retained or passed to the coalition commander:

- **Role 1A.** The JTFC retains operational command of the assigned force.

- **Role 1B.** The JTFC retains operational command but transfers operational control of the assigned forces to another headquarters.

- **Role 2.** The JTFC transfers operational command of the assigned forces but remains as the Canadian National Commander.[30]

The project team developed three progressive operational-capability levels in line with the roles of the JOG. Operational-capability Level 1 was the development of some key capabilities in support of contingency operations, such as the command and control of humanitarian operations, the provision of operational reconnaissance teams, and activation and closeout of a theatre of operations. This *initial operational capability* was reached in October 2000. Operational-capability Level 2 built on this and included the training and development of operational-level staff to allow the joint force commander to effectively command and control CF elements as part of a coalition led by another nation. This *final operational capability* was reached on 20 June 2003.[31] A further operational capability, never realized, was to be the development of additional staff and operational-level capabilities so that Canada could be the lead nation in a coalition. The relationship of the operational capabilities, the types of operations, and the assigned roles is shown at Table 5.3.[32]

The three operations that will be examined are in bold in Table 5.3. The first, Operation Abacus, was a good example of a Role 1A operation. Operation Apollo and Operation Eclipse were good examples of Role 1B and Role 2, respectively. This review of the operations is based on primary source documents, after-action reports and lessons-learned staff action directives.

	Initial Operational Capability 15 Oct 2000		Final Operational Capability 20 Jun 2003		Not Reached
	Operational Capablity Level 1	Operational Capablity Level 2			Operational Capablity Level 3
Type of Operation and Command Relationships	Domestic & International Operations Functioning as Op Commander (DART HQ) Assistance to Op Commander (Ln, Recce, Theatre Activation and Mission Closeout)	Domestic Operation Functioning as Op Commander (JTFHQ) Retain Operational Command Retain Operational Control	International Operation Not Functioning as Op Commander (JTFHQ/NCE) Retain Operational Command Pass Operational Control	Pass Operational Command Retain Admin. Control	International Operation Functioning as Op Commander (CJTFHQ) Retain Operational Command Retain Operational Control
	Partial Capability All Roles	Role 1A	Role 1B	Role 2B	Canada as the lead nation (Role 1A)
Exercises and Operations	OP CENTRAL (1998) OP TORRENT (1999) Ex JOINT ACTIVATION (2000) OP FORAGE (2001)	OP ASSISTANCE (1997) OP RECUPERATION (1998) OP ABACUS (1998-2000) Ex JOINT WOLF (2002)	Ex UNIFIED SPIRIT (2000) OP ECLIPSE (2000) Ex JOINT JAVELIN THRUST (2001) OP APPOLLO (2001-2003) Ex COOPERATIVE JAGUAR (2003)		OP ASSURANCE (1995)

TABLE 5.3. CF JOINT OPERATIONS GROUP COMMAND RELATIONSHIPS, ROLES AND EXERCISES/OPERATIONS

SOURCE: ADAPTED FROM J.P.Y.D GOSSELIN, "CF JOG FULL OPERATIONAL CAPABILITY" CANADIAN FORCES JOINT OPERATIONS GROUP, KINGSTON, FILE NUMBER 1901-2 (J5 MAR) DATED 22 APRIL 2003.

Operation Abacus. Operation Abacus was an example of a contingency operation where the employment of operational-level headquarters proved to be highly successful. At the time, NDHQ was being stretched in maintaining control of operations around the world. The CDS recognized that the scope of the Year 2000 problem was well beyond the capability of CF to handle alone, and that NDHQ staffs, led by the DCDS, would need to be engaged with other government departments and other national militaries in preparing for potential problems prior to, during and after midnight on 31 December 1999. Therefore, it was agreed that an operational-level commander would be required to command and control the CF in Canada while the DCDS would command and control the CF deployed overseas. As a result, the commander of the 1st Canadian Division was appointed the JTFC for the operation. The operation formally commenced in March 1998 and ceased in February 2000.

Due to the operational tempo of the CF at the time,[33] the DCDS tasked the JTFC with drafting the strategic-level guidance and plan. The approved Strategic Planning Guidance clearly stated the split of responsibilities between the staffs. The DCDS was responsible for strategic command and control (C^2), intelligence, co-ordination of resources, and direction. The JTFC was given the responsibility for operational-level planning and execution of the operation.[34] The joint task force headquarters used the operational planning process extensively.[35]

The command and control structure was based on four subordinate joint task forces and a Joint Force Air Component Headquarters co-located with the JTF headquarters. The command and control structure was successful because it was based on a clear chain of command with clearly identified commanders at all levels of command. The command relationships between the various elements of the CF were practised through several major exercises until they were well understood.

The various exercises in preparation for the operation also brought to light a number of areas that needed improvement. There was misunderstanding of basic C^2 terminology among the Environmental Chiefs of Staff (ECS) and NDHQ staff. Familiarity with joint doctrine varied throughout the levels of command, and concerted effort was made to bring staffs to the same level.[36] Due to the complexity of the operation, the DCDS Instruction for Domestic Operations was unsuitable and had to be rewritten.[37]

Overall, both the preparations and the actual operations were highly successful. The fact that there were dedicated staffs at the operational level to address these issues meant that they could be resolved over time without great impact on the strategic-level staff, which continued to control contingency operations overseas. With the formation of the JOG six months later, it was anticipated that future operations would follow the same approach. Unfortunately, that was not the case.

Operation Eclipse. In the summer of 2000, while the JOG was involved in training new staff and preparing to achieve its initial operational capability, the war between Eritrea and Ethiopia was reaching a climax and the UN was trying to deploy a peacekeeping force in the region. Canada, along with the Netherlands, had previously agreed to contribute forces to the United Nations' Standby High Readiness Brigade. This was an opportunity to introduce the JOG to operations, and it was also an opportunity to practise the JOG's newly developed theatre activation team and mission-closeout capabilities. The operation formally commenced on 6 December 2000 with the arrival of the theatre activation team and ended with the departure of the mission-closeout team in July 2001.[38]

The CF commenced strategic planning in conjunction with the Netherlands, and it was agreed that Canada would provide an infantry company group (from Second Battalion, Royal Canadian Regiment) as part of a Dutch infantry battalion. National Defence Headquarters decided not to involve staffs outside of Ottawa, including the JOG, due to concerns over the release of the information that the Netherlands was considering joining Canada in the operation, before its government approved that decision. The DCDS staffs, who were already overworked, were now required to conduct both strategic and operational planning for another contingency operation.

In order to save time, the formal operational planning process was not followed, which caused many problems later on. Though the information was available to do so, the CDS did not produce an initiating directive. While some staff planning guidance was provided for the mission analysis phase, in subsequent planning phases only verbal guidance was offered.[39] No written guidance was ever produced for the company group option, and there was never a statement of the commander's intent from the strategic level. The Land Staff actually wrote the strategic mission statement for subsequent approval by the J Staff. Insufficient information was available to allow the staff to develop any formal course of action analysis or recommendations. Throughout the planning period various members of the J staff made different assumptions, occasionally working at cross-purposes. This lack of a common understanding and unity of purpose resulted in staff time and effort being wasted.[40]

The Joint Operations Group was tasked with providing the theatre activation team for the operation. The group developed a campaign plan, but attempts to co-ordinate the JOG campaign plan with the strategic plan were unsuccessful. As a result, the theatre activation team deployed into theatre without clear strategic and operational objectives. Due to the late appointment of the Canadian national commander, the commanding officer of the theatre activation team became the de facto national commander during the early part of the mission. The theatre activation team passed control of the operation to the national command element staff after seven weeks. Six months later, at the end of the operation, the JOG provided a mission-closeout team to allow the CF elements to depart quickly out of the theatre back to Canada.

In summary, the operation raised many issues related to the role of the JOG in contingency operations. The late appointment of the Task Force Commander affected unity of command. The lack of unity of purpose in the NDHQ joint staff resulted in uncoordinated staff action. The fact that the JOG was not included in the earlier planning and that later the NDHQ staff were unwilling to review their campaign plan made it difficult to establish trust and mutual understanding between the staffs. Though the deployments of the theatre activation team and the mission-closeout team were very successful, the JOG had yet to prove itself in the command of a mission. That opportunity was to follow a short three months later.

Operation Apollo. Soon after the terrorist attacks on the United States on 11 September 2001, the CF started discussing with the United States options for forces that Canada could contribute to the fight against terrorism. On 9 October, a reconnaissance team travelled from Ottawa to the US Central Command in Tampa, Florida. Over approximately the next two weeks, the team conducted strategic planning to determine Canada's contribution. On 26 October, an operational-level formation was established known as the Canadian Joint Task Force South West Asia, and the staffs of the JOG were used to form the JTF headquarters. At its peak, the Canadian contribution included a Canadian naval task group of four warships; an infantry battle group from Third Battalion, Princess Patricia's Canadian Light Infantry; strategic and tactical airlift; long-range patrol aircraft; and a National Support Unit (NSU).

The split of responsibilities between the JTF and NDHQ established a clear chain of command. The JTFC was made responsible to the DCDS for all matters related to national command, including the operational readiness, administration and discipline of the task force. He was also responsible for monitoring the operational employment of the JTF; taking necessary action to ensure Canadian policies were respected; conducting liaison with Commander Central Command; and ensuring that the DCDS was informed of significant issues.[41]

The command and control structure at the operational level was complicated because the operational headquarters was in Tampa while the JTF units consisted of differing types of units in a variety of geographic locations throughout southwest Asia. The JTFC made several requests to NDHQ for permission to deploy his headquarters into the theatre in order to ensure effective command, but they were all refused because of the need by NDHQ to have access to information from the Central Command headquarters. As a result the commander spent much of his time travelling to and from the theatre. The JTFC for Rotation 1 considered that the separation of the deployed elements from the operational-level commander was a risky command structure.[42]

There were a number of planning and control issues that arose during the operation. The strategic planning team in Tampa focused its efforts on providing whatever forces were available based on their readiness levels and without fully considering the implications with respect to sustainment in the operational planning process.[43] Consequently, all the support elements were task tailored for a specific capability, resulting in six different support elements in the theatre at different locations.[44]

It was also difficult to conduct medium-term planning for the ongoing operation.[45] While the operational-level planners in Tampa worked with Central Command staff, NDHQ had no strategic-level mid-term planners. It had to reassign J3 staff to assist the J5 staff in strategic planning tasks, while the J4 staff did all the strategic and operational planning within NDHQ. This created an imbalance of J3-J4-J5 planning responsibilities in NDHQ and between NDHQ and the JTF.

There was constant pressure by NDHQ to get information from the theatre on the activities of the JTF; however, the JTF headquarters could seldom filter requests or provide the information requested by NDHQ. As a result, NDHQ often bypassed the chain of command and contacted the units directly. Answering these requests required considerable effort for the units, especially the infantry battalion, and created a separate flow of information outside of the chain of command.

In summary, Operation Apollo demonstrated again the need to have an operational-level headquarters, like the JOG, available to conduct planning and control of the assigned forces. The inability of the JTFC to deploy into the theatre resulted in command and control problems, and the demand by NDHQ for information from theatre resulted in its violation of the chain of command. Furthermore, the NDHQ staffs were inadequately structured and resourced to conduct their strategic planning functions.

Options and a Solution

> There is nothing more difficult to take in hand, more perilous to conduct or more uncertain in its success, than to take the lead in the introduction of a new order of things.
>
> Machiavelli[46]

The ability of the JOG to contribute to the success of CF operations over such a short period of time demonstrated the value of having a well-trained operational-level headquarters. Success was achieved when there was a clear split in responsibilities between the strategic and operational levels, the chain of command was respected and authority was decentralized to allow for freedom of action. Problems occurred when there was a lack of unity of purpose due to unclear strategic direction, incomplete campaign planning, and lack of co-ordination between operations and support staffs at the strategic level.

A comparison of the actions of both NDHQ and the JOG on the above operations against the proposed CF principles of command is conducted at Table 5.4. The results suggest that when the principles of command are followed, the operation is more successful in terms of command and control. The merging at NDHQ of strategic and operational functions needed to be re-examined with the arrival of the JOG on the scene. There was a requirement to structure the staffs in NDHQ so that they could achieve both unity of command and purpose. In terms of planning, the strategic level should have focussed on developing military strategy, while the operational level carried on with campaign planning. In both cases, the formal operational planning process should have been followed. During the conduct of operations, NDHQ should have used the operational-level headquarters to provide them information instead of contacting the tactical units directly. Clearly, there was a need for change as there were constant problems with command and control on CF operations. The question is, what were the options available at the time to resolve the situation?

Principles	Operation Abacus	Operation Eclipse	Operation Apollo
Unity of Command	Yes	No. The JTF Commander was appointed too late to deploy on the strategic reconnaissance or arrive in theatre with the theatre activation team.	Yes
Chain of Command	Yes	Yes	No. NDHQ violated the chain of command to get information on the task force elements directly
Decentralized Authority	Yes	Yes	No. The task force commander was not allowed to deploy his headquarters into the theatre of operations to effectively command the task force

...continued on next page

Principles	Operation Abacus	Operation Eclipse	Operation Apollo
Span of Control	Yes	Yes	Yes
Unity of Purpose	Yes	No. The incomplete use of the OPP caused a lack of unity of purpose for many of the NDHQ staffs, which resulted in problems in co-ordination.	No. The lack of agreemeent on producing a co-ordinated sustain-ment plan resulted in overlaps in sustainment efforts until the NSU was created.
Trust and Mutual Understanding	No. Understanding of joint doctrine varied greatly.	No. Trust between the JOG and DCDS staffs was strained due to the JOG's exclusion from the early planning.	Yes

TABLE 5.4. COMPARISON OF COMMAND PRINCIPLES AND OPERATIONS

The options for resolving these issues are related to how best to assign between NDHQ and the JOG the responsibility for strategic- and operational-level command and control functions. During the period examined by this chapter, NDHQ performed many of the operational functions related to contingency operations. On the other hand, the JOG performed only some of the operational functions in peace, such as non-combatant evacuation,[47] while it was negotiating the transfer of others, like the CF Contractor Augmentation Program.[48] Any solution should have respected the principles of command, clarified what and where strategic- and operational-level command and control functions would be performed, and also addressed the JOG's ability to provide effective command and control during peace and war.

Two options that could have been chosen are described here. The first option, called NDHQ Plus, gives the DCDS the responsibility for both strategic and operational functions while retaining the JOG for large domestic and international operations. The second option, called JOG Plus, defines and allocates responsibilities so that NDHQ would have been responsible for strategic-level functions and the JOG for operational-level functions.

The NDHQ Plus option has been considered before. In 2000, the Vice-Chief of Defence Staff contracted Vice-Admiral (Retired) Mason and

Lieutenant-General (Retired) Crabbe to complete a study of creating a single centralized, operational-level HQ for the CF. They concluded that such an HQ model not be adopted but that a new "third" option be considered in its place. This option would transfer all force employment responsibilities to a single operational-level CF joint headquarters under command of the DCDS. The NDHQ Plus option is similar, but it would only retain the JOG for certain contingency operations that require the deployment of a headquarters.

The advantages of the NDHQ Plus option are that the ability to provide information to the CDS and government would have been enhanced for most contingency operations; and command and control would have been centralized in the DCDS except when there was a need for a deployable headquarters. The disadvantages are that the distinction between the strategic and operational planning functions would have been difficult to maintain, and the structure would not have been the same for crisis and war. Moreover, the investment made in forming the Joint Operations Group and the Joint Support Group would not have been fully realized.

In terms of principles of command this option provides some concerns. Unity of command and a proper chain of command would have been difficult to achieve as both staffs would have been reporting to the same commander. If the JOG had been tasked to deploy a headquarters, a new chain of command would have had to have been established. Unity of purpose, trust and mutual understanding would have been enhanced between the strategic and operational staffs due to their centralization in a single organization, but during operations they would not have been enhanced between the staff in NDHQ and the tactical units deployed on operations. This option would not have allowed for decentralized decision making or allow for freedom of action at the tactical level.

The JOG Plus option is based on passing to the JOG the responsibility of operational-level command and control for all contingency operations. The JOG would also have maintained a deployable headquarters capability for contingency operations, while NDHQ would have retained the responsibility for strategic-level functions. This option is based on the United Kingdom's Permanent Joint Headquarters (PJHQ) model.

In 1996, the United Kingdom created a single permanent joint headquarters to permit a clear connection between government policy and strategic functions and the conduct of operations at the operational level.[49] It is headed by a three-star commander, Chief of Joint Operations, who is responsible to direct, deploy, sustain and recover all UK forces deployed on operations. A permanent staff of 438 personnel organized along the continental staff system supports him. The staff is divided into operations and support staffs, each commanded by a two-star general. Under command of the two-star general responsible for operations is a deployable joint headquarters, similar to the JOG, commanded by an Army or Royal Marine brigadier, and consisting of a staff of 60 personnel. A dedicated communications squadron, intelligence battalion and pioneer platoon support the joint headquarters, and it is capable of forming a JTF headquarters or national command element.

In terms of split of responsibilities, the British minister of defence provides strategic guidance and direction, and the PJHQ focuses on campaign planning and operational command of deployed forces. The PJHQ commander is often appointed as the operational commander for contingency operations, which allows for an effective command-driven approach. There is a highly organized mid- and long-term planning process that allows for easy transfer of responsibilities to operations staffs. The equal representation of operations and support issues ensures that the operations are effective and sustainable.

The JOG Plus option would have followed this model. The Joint Operations Group would have continued to be in the DCDS's chain of command for force generation and force employment issues; however, the JOG would have been given the authority to co-ordinate with the other eight operational-level headquarters to facilitate domestic operations and to standardize operational-level doctrine and procedures. The organizational changes would have been done by first determining strategic and operational functions and then adjusting resources between the DCDS and the JOG. The aim would have been to eliminate any duplication between JOG and NDHQ joint staffs. A new staff appointment of Chief of Staff J4 would have been created at NDHQ and the JOG to co-ordinate all support functions, including personnel, logistics, finance and legal. Communications and information systems (J6) staffs would have been moved under the COS J3 at both headquarters. The commander of the JOG would have been a general officer, and the current JOG commander would have been designated his chief of staff. The control of contingency operations would have moved from the National Defence Command Centre (NDCC) to the JOG, and NDCC would have focussed on co-ordination of military capabilities with other governments and militaries. The communication and information capabilities at the JOG headquarters in Kingston would have needed to be examined and improved, as required. This would have also permitted the JOG headquarters to become the alternate command post for National Defence in the case of a crisis or war.

The advantages of this option were that there would have been a clear split in responsibilities between the strategic and operational levels, the structure would be the same in peace, crisis and war, and it could easily have been used to effect control of contingency operations. In addition, this option would have allowed for the deployment of the joint headquarters while maintaining the permanent staff in Kingston to continue to command other contingency operations. The disadvantage of this option was that it would have required close co-operation between the strategic and operational staffs in preparing for operations.

In terms of the principles of command, this option provides unity of command throughout peace, crisis and war. It would have allowed for the early appointment of a joint task force commander, and it would have reduced the violations of the chain of command as there would have been a hierarchical and clear command structure that would have been consistent with NATO doctrine. This option also supports decentralized authority and eliminates information flows from tactical directly to strategic levels. There would also have been a clear continuity of command, especially when higher-level C^2 systems were disrupted. It would have enhanced unity of purpose at and between the strategic and operational levels with the re-establishment of equal consideration of operations and support functions. Furthermore, this option would have improved trust and mutual understanding at all levels within the CF, and it would also have enhanced mutual understanding with allies and partners as it was based on approved and commonly used command and control doctrine.

In comparing the two options, it is clear that the JOG Plus option is far superior to the NDHQ Plus option, both in terms of how it addresses the concerns raised on the operations described above and how it meets the principles of command. The changes that should have been implemented were not great, but the rewards were potentially large and attractive. Much time and effort were spent on creating the JOG and its subordinate formations and units. Yet, the allocations of responsibilities and resources did not meet the requirements for CF contingency operations at the time.

Conclusions

This chapter has argued that that when the JOG had reached its full operational capability, it should have been given the full authority and responsibilities for the command and control of all CF contingency operations, which were, at that time, conducted at the strategic level by NDHQ. A detailed examination of CF concepts and doctrine in the areas of command and control responsibilities, planning and execution exposes the discontinuities between NATO and CF doctrine in these areas. Through the use of after-action reports and lessons-learned directives, three CF operations were explored in terms of command and control at the strategic and operational levels. The results were then compared against enhanced CF principles of command. From this analysis a number of criteria were presented along with two potential options that would have resolved these command and control concerns with contingency operations. The first option, NDHQ Plus, would have moved the operational-level command and control functions to NDHQ in order to meet demands for centralized command and control of contingency operations. It was shown that this option was unworkable as it does not support many of the command principles and is not consistent for peace, crisis and war situations. The second option was based on the United Kingdom's Permanent Joint Headquarters model. It proposed making the JOG responsible for the operational-level command and control of all contingency operations. It was found to be the best option as it clearly splits strategic and operational responsibilities, and it would have allowed for easy, flexible command and control of contingency operations whether in peace, crisis or war.

NOTES

1   Sun Tzu, *The Art of Warfare*, Roger T. Ames, trans. (New York: Ballantine Books, 1993), 113.

2   The Defence White Paper directed a reduction of one third in the resources devoted to headquarters' functions. This cut was later increased to one half by NDHQ. Canada, Department of National Defence (DND), *1994 White Paper* (Ottawa: Canada Communication Group, 1994), 41.

3   G.L. Garnett, "The Evolution of the Canadian Approach to Joint and Combined Operations at the Strategic and Operational Level," *Canadian Military Journal* 3, no. 4 (Winter 2002), 4.

4   Jean H. Morin and Richard Gimblett, *Operation Friction, 1990–1991: The Canadian Forces in the Persian Gulf* (Toronto: Dundurn Press, 1997), 113–26. The challenges faced in forming a joint headquarters on operations were not lost on NDHQ staffs. There was a clear need for a deployable joint HQ.

5   DND, "Development of Canadian Forces Joint Operational Level Command and Control Capability," NDHQ Action Directive D/12/94 dated 29 October 1994, 1.

6   General J.E.J. Boyle, "Joint Operational Level Command and Control Capability," NDHQ Action Directive D/3/96 dated May 1996, para. 8, p. 7.

7   These operations included support to Manitoba due to floods in 1997; support to Ontario, Quebec and New Brunswick due to the effects of an ice storm in 1998; and the potential support to the federal government and all provincial governments in December 1999 to January 2000.

8   DND, *CF Obtains New Capability: A Deployable Joint Headquarters*, http://www.forces.gc.ca/site/operations/CFJOG/article_e.asp (accessed 16 September 2003).

9   Carl Von Clausewitz, *On War*, Michael Howard and Peter Paret, eds. and trans. (Princeton, NJ: Princeton Univ. Press, 1976), 141.

10   The eight headquarters are the two naval headquarters on the west and east coasts (MARPAC and MARLANT, respectively); the four Land Force area headquarters in Edmonton, Toronto, Quebec City and Halifax; the 1 Canadian Air Division in Winnipeg; and CF Northern Area headquarters in Yellowknife.

11   DND, *Canadian Forces Operational Planning Process*, B-GJ-005-500/FP-000 (6 November 2002), 1–3.

12   DND, *1994 White Paper.*

13   DND, *Shaping the Future of the CF: A Strategy for 2020* (1999), 6–10.

14   Integrating concepts include *network-enabled operations* (NEOps); *joint, interagency, multi-national and public* (JIMP); and *effects-based operations* (EBO).

15   Functional concepts include *human resources, information management,* and *sustainment.*

16   DND, *Strategic Operating Concept (Draft)*, Supporting Concepts – C2, http://www.vcds.forces.gc.ca/dgsp/pubs/rep-pub/dda/cfsoc/chp4_e.asp (accessed 7 October 2003).

17   DND, "CF Joint Operating Concept 2012," draft dated 24 July 2003, 1–3.

18   The continental staff system is used throughout the world to organize staffs. The Canadian system has nine branches: J1, personnel; J2, intelligence; J3, operations; J4, logistics; J5, plans; J6, communications and electronics; J7, training and doctrine; J8, finance; and J9 civil military co-operation.

19   See DND, DCDS Group, J7 Joint Doctrine Web Site, Document Hierarchy, http://www.dcds.forces.gc.ca/jointDoc/pages/j7doc_doclist_e.asp (accessed 12 October 2005).

20   NATO, *NATO Glossary of Terms and Definitions*, AAP-6 (English and French) (Brussels: NATO Standardization Agency, December 2002), p. 2-S-10.

21   Ibid., p. 2-O-2.

22   NATO, *Allied Joint Doctrine*, AJP-01 (A) Change 1 (Brussels: Military Agency for Standardization, April 1999), 20.

23   DND, *Canadian Forces Doctrine (Third Draft)*, 14.

24   Ibid., 42–3.

25   DND, *Canadian Forces Operations*, 3–1.

26   DND, *Canadian Forces Doctrine (Third Draft)*, 66.

27   Ibid., 62.

28   Van Creveld, *Command in War*, 15.

29   DND, DCDS, Project Charter DSP 2001, CF Joint Headquarters / Joint Task Force Headquarters, Version AL-1, dated 15 November 2000, 3–4.

30   DND, DCDS, Project Charter DSP 2001, CF Joint Headquarters / Joint Task Force Headquarters, dated March 1999, 2–3.

31   DND, *CANFORGEN 082/03 CDS 074 201945Z Jun 03 CF Joint Operations Group (CF JOG) Full Operational Capability (FOC) Declaration*, http://vcds.dwan.ca/vcds_exec/pubs/canforgen/2003/082-03_e.asp, DWAN (accessed 1 October 2003).

32   Adapted from J.P.Y.D Gosselin, "CF JOG Full Operational Capability," Canadian Forces Joint Operations Group, Kingston, file number 1901-2 (J5 Mar) dated 22 April 2003, 12.

33   During the period March 1998 to January 2000, the CF was involved in conflict and, later, peace-keeping in Kosovo and in UN peacekeeping in East Timor, in addition to Bosnia.

34   Lieutenant-General R.R Crabbe, "Strategic Direction Operation Abacus (Y2K)," DCDS file number 3000-15, dated 31 August 1998, Annex B.

35   Major D. Burden, "NDHQ Joint Staff J3 Lessons Learned Questionnaire OP ABACUS," dated 6 December 1999 [sic], 18.

36   Ibid, serial 10, 5.

37   Ibid., serial 8, 4.

38   Major General J.O.M. Maisonneuve, "Operation ECLIPSE: Lessons-Learned Staff Action Directive (LLSAD)," NDHQ file number 3350-165/E9 (J7 Lessons Learned), p. A1.

39   DND, "DCDS Joint Staff Planning Guidance: Options for UNMEE," NDHQ file number 3350-1 (J3 Intl 2-1), dated 6 June 2000.

40   Maisonneuve, "Operation ECLIPSE: Lessons-Learned Staff Action Directive (LLSAD)," p. A3/15.

41   Colonel J.P.Y.D. Gosselin, "Joint Force Command and Control," briefing to CSC Course 29, 21 November 2002, slides 20, 22, 24 and 26.

42   Major-General J.S Lucas, "Operation Apollo: Lessons-Learned Staff Action Directive (SAD)," NDHQ file number 3350-165/A27 (J7 Lessons Learned) dated 30 April 2003, pp. B-14 to B-15.

43   Ibid., pp. B-6 and B-38-39.

44   Ibid., p. B-38. There was one support element for the navy, four support elements for the air force and one for the army.

45   See Commander CJTFSWA OP APOLLO ROTO 1 "Tour End Report," file number 3350-134-1 (Comd) dated 21 October 2002, Annex A, para. 47.

46   Niccolo Machiavelli, *The Prince*, trans. George Bull (London: Penguin Books, 2003), 21.

47   Commodore J.J. Gauvin, "Non-Combat Evacuation Operation (NEO): Joint Operation [sic] Group (JOG) Way Ahead," NDHQ file number 3450-7 (J3 Intl Plans 2), dated 19 March 2003, para. 1.

48   Colonel J.M.Turner, "CF JOG Reservations, First Edition of CANCAP Program Governance Document: Devolution of Oversight Responsibilities at the Operational Level," Canadian Forces Joint Operations Group, Kingston, file number 1901-1-10 (Comd), dated 11 July 2003, 1–2.

49   Richard M. Connaughton, "Organizing British Joint Rapid Reaction Forces," *Joint Force Quarterly* no. 26 (Autumn 2000), 88.

CHAPTER 6

OPERATIONAL-LEVEL LEADERSHIP AND COMMAND IN THE CANADIAN FORCES: GENERAL HENAULT AND THE DCDS GROUP AT THE BEGINNING OF THE "NEW WORLD ORDER"[1]

General R.R. Henault, Brigadier-General (Retired) Joe Sharpe, and Allan English

The fall of the Berlin Wall in November 1989 marked, for some, the end of the Cold War and the beginning of a "new world order." Other events in the last decade of the twentieth century also marked the transition to a new world order in which the Canadian Forces began to conduct a whole new range of operations in addition to those they had previously conducted. These new operations, from the high-intensity military operations of the 1991 Persian Gulf War to large-scale aid to the civil power in the 1997 Manitoba floods, required an increasingly joint response from the CF, where land, air and maritime forces worked in unison to achieve a common aim. The operations were conducted some-times in an alliance or coalition context and sometimes in a domestic context, but whatever the context, they required new command and control structures to ensure their success.

In the last decade of the twentieth century, the command and control of CF joint and combined operations started an evolution from control structures that were largely single environment–based[2] to a structure where command and control (C^2) was exercised by the Deputy Chief of the Defence Staff (DCDS) on behalf of the Chief of the Defence Staff (CDS). During that decade, the DCDS Group at NDHQ in Ottawa was instrumental in creating new C^2 frameworks to meet the challenges posed by these new operations. General Raymond Henault was a key player in the creation of the new frameworks, and in many ways he defined the role of DCDS in the new world order because he served in essentially the DCDS role for four years, during the period from 1996 to 2001, except for one year in 1997–1998 when he was Assistant Chief of the Air Staff. This was, as we shall see, a critical period in the evolution of joint C^2 in

Canada, and Henault continued to be involved in its evolution from June 2001 to November 2004 when he served as CDS. He continues to follow issues of joint and combined C^2 with great interest in his position as Chairman of the NATO Military Committee in NATO Headquarters in Brussels, which he assumed in June 2005.

Before examining General Henault's role in the evolution of CF operational-level C^2, it is necessary to put his actions in the context of the times. The first section of this chapter describes the beginning of the post–Cold War world and the new world order of the day.

Context

The end of the Cold War ushered in many changes, including the 1989 announcement by Warsaw Pact nations of deep cuts to their conventional forces, the unification of Germany in October 1990 and the break-up of the Soviet Union in December 1991. For most Western governments this "new world order" brought expectations of a peace dividend, and Canada, like its allies, began to reduce its military forces in the early 1990s.

The downsizing of the CF in the 1990s was precipitated by defence budget cuts as part of the government's deficit-reduction program.[3] The CF's contribution to deficit reduction was achieved by cutting the defence budget, which totalled $12 billion in 1993–1994, to $9.38 billion in 1998–1999.[4] Unfortunately for the CF, Treasury Board adopted a methodology for making cuts to the Department of National Defence (DND)[5] that eschewed careful consideration of how cuts to the CF could be achieved while maintaining its operational effectiveness. Instead, resources were removed from DND, making the status quo unsustainable, and then the CF was forced to scramble to make cuts while still conducting operations.[6] Because many of the budget cuts made in the 1990s were unforecast, this caused chaos in the CF's downsizing program.[7]

The cumulative effect of the cuts and the government policies devised to implement them had a serious impact on the ability of the CF to conduct its roles and missions in an effective and sustainable fashion at the beginning of the twenty-first century. As a result of post–Cold War budget cuts, the CF was reduced from about 90,000 regular force

personnel in 1990 to approximately 62,000 regular force personnel in
2005.[8] Today's force size reflects a 25-year low for the CF. Much of the
personnel reduction was carried out under the Forces Reduction
Program, which was created in 1992 to reduce the complement of the CF
by encouraging members to take early retirement. The plan continued
until the end of the 1997–1998 fiscal year.[9]

At the same time that significant cuts were being made in CF personnel
strength and budget allocation, government demands for CF participation
in various operations increased. Instead of a relatively quiet post–Cold
War world where governments could reap a peace dividend, a new world
disorder greeted policy-makers. Starting with the first Gulf War in 1991,
continuing through a variety of crises precipitated by failed states, and
culminating in the post–11 September 2001 "war on terror," this anarchic
international situation produced a number of scenarios in which the
Canadian government decided that it wanted to intervene with military
contributions. Combined with demands for CF participation in a number
of domestic operations during this period, on average the CF was far
busier than it had ever been during the Cold War.

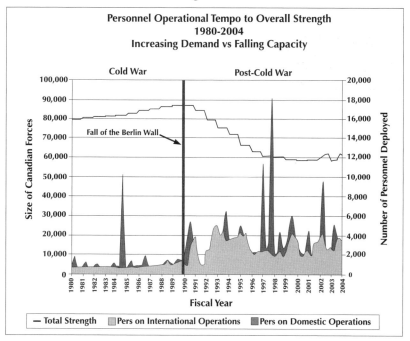

FIGURE 6.1. PERSONNEL OPERATIONAL TEMPO TO OVERALL STRENGTH,
1980–2002[10]

Figure 6.1 illustrates how the increase in the number of CF deployments has coincided with a decrease in the number of CF personnel available for service during the post–Cold War period. While the CF's resources (budget and personnel strength) were cut by about 20 percent, the number of its personnel deployed on operations increased threefold.

In the post–Cold War period the CF has participated in two major categories of operations: routine and contingency. Routine operations take place on a regular basis, and forces are specifically tasked, organized and equipped for these pre-planned operations. Contingency operations tend to be launched in reaction to a crisis or a natural disaster, and forces are generated as required to meet the specific needs of every mission. Both routine and contingency operations can take place in either a domestic or an international context. Domestic contingency operations usually consist of the CF providing aid to the civil power, while for international contingency operations CF missions are initiated by the government in support of its foreign policy objectives.

The most visible and most publicized operations conducted by the CF are "crisis" international contingency operations that are mounted in response to an international crisis or natural disaster. In addition to crisis contingency operations, since the end of the Cold War the CF has conducted standing and continuous commitments for the North Atlantic Treaty Organization (NATO), for the North American Aerospace Defence Command, and in the Balkans. Large forces, by Canadian standards, have also been deployed to southwest Asia, the Middle East, Bosnia and Africa. The CF has participated in approximately 20 international contingency operations since the end of the Cold War, as illustrated in Figure 6.2. The number and scope of these operations have created a high operational tempo for the CF as large numbers of CF personnel have been employed in post–Cold War routine and contingency operations both at home and abroad.[11]

FIGURE 6.2. INTERNATIONAL CONTINGENCY OPERATIONS, 1990-2004[12]

FIGURE 6.3. DOMESTIC CONTINGENCY OPERATIONS, 1990–2004[13]

When Henault started working in the DCDS Group, there was a limited joint-staff capability, and the command and control of many operations was carried out in an ad-hoc manner under the direction of the Environmental commanders. The Army, Navy and Air Force, each commanded by a three-star general, were operating for the most part independently of each other and of NDHQ from their headquarters in St Hubert (Quebec), Halifax and Winnipeg, respectively. Each Environment had a representative at the one-star level in Ottawa — for example, the Director General of Air Doctrine and Operations (DG ADO) represented the Air Force — with the commanders of the Environments coming to Ottawa once a month to consult with NDHQ staff and attend meetings of the Defence Management Committee and the Force Development Steering Group (the major resource-allocation committee). Therefore, the commanders' knowledge and involvement in what was going on in NDHQ was relatively low, particularly with respect to domestic operations. Likewise, the lack of close contact with the Environmental staffs also meant that members of the joint staff did not have the information they required to do their jobs effectively. For example, during the 1991 Gulf War, when asked questions about Army operations by the politicians or their staffs, the fledgling joint staff found they could not answer the queries except to say that they would have to go check with the Army to get the answers and get back to their questioners a week later. This example of the lack of clarity in command relationships also illustrates the confusion that existed between force employment and force generation at the time.

One of the key issues that were resolved during Henault's time in the DCDS Group was the distinction between force employment and force generation — effectively, the issue of who would command Canadian joint forces on operations. In the early 1990s the issue had not been satisfactorily resolved, and the commanders of the Army, Navy and Air Force actually commanded their forces when they were on operations. The DCDS could give guidance or ask for support, but he had no direct force employment or command responsibilities when Henault came to the DCDS Group in 1996. During the 1991 Gulf War and the Oka Crisis, this "sharing" of responsibility for force employment created tremendous confusion within the CF.

General Henault's Experiences

Before arriving at NDHQ to take over as Chief of Staff, Joint Ops (COS J3), for the DCDS Group, Henault had been Commander of 10 Tactical Air Group from 1994 to 1995 and then Chief of Staff Operations in Air Commander Headquarters in Winnipeg from 1995 to 1996. He was asked in early 1996 by the Chief of the Defence Staff to come to NDHQ to take the COS J3 job. Henault believes that his previous appointments plus his one-year National Defence College course had prepared him well for his new role. However, he only spent about six months as COS J3 before he became Acting DCDS. In fact, when he arrived in Ottawa, after about a one-month period of hand-over, he became Acting DCDS because the DCDS was heading off on his summer leave.

Henault had a quick introduction to his new job as he started work on a Friday and, that same weekend, the Saguenay floods occurred. This was his introduction to being not only COS J3 but also Acting DCDS dealing with the response and the aftermath of the Quebec floods. The Saguenay floods in 1996 were the first major domestic operation since the Oka Crisis (Operation Salon) in 1990, where a concerted military response was required in aid to the civil power. In hindsight, compared to other operations since, the CF Saguenay flood response (Operation Saguenay) was not that large, but it was nonetheless a significant response to an emergency event, with the CF base at Bagotville really being the focal point for the response. From Henault's perspective, Operation Saguenay, using an area concept for domestic operations as a foundation for regional response activities, was really the beginning of the domestic operations philosophy that the CF ultimately adopted.

Operation Saguenay was co-ordinated by desk officers in Ottawa at NDHQ through what became the *J3 Continental Desk*, essentially the domestic operations desk. It was realized in retrospect that this was not a particularly robust command and control framework as there was only a very rudimentary J Staff at that stage and no real joint-staff mechanisms like the Crisis Action Team that existed until recently. Until reforms were made in the joint-staff structure, there was no effective way to link into the other elements of DND, especially the support, intelligence, medical and communications elements with which the CF deploys, as the Crisis

Action Team did until recently. Therefore, even though the CF response in Operation Saguenay was relatively good, it did not have the joint-staff oversight that the CF has now.

The lack of a proper joint staff required Henault to take a very hands-on approach as the Acting DCDS during Operation Saguenay. His acting COS J3 also was actively involved in the operation, but the lack of joint-staff capability meant that the base commander in Bagotville was responsible for a great deal of the response. He had a direct link into the area head-quarters in Montreal and a direct link to NDHQ to try and co-ordinate his requirements. The co-ordination of those requirements was very difficult for the base commander because he did not have a proper command and control structure at his base to respond to all of his needs. In fact, even getting his support needs required him to go to several different agencies. So, looking at the situation from the perspective of the commander in the field, he did not have adequate command and control support from higher headquarters. Despite these shortcomings, the operation was a success, and much of the credit for that success goes to the base commander and his team in Bagotville.

Operation Saguenay was a turning point, for Henault, in the evolution of the CF joint C^2 framework. It was recognized that the method of co-ordinating the activities of the Army, Navy and Air Force through desk officers in NDHQ was not an effective C^2 framework for joint operations because there were no clearly defined lines of command, authority and responsibility. For example, even though he had a reasonable amount of embedded communications capability in the DCDS Group, Henault had great difficulty co-ordinating actions of commanders in the field, ships and aircraft detachments because, with parallel but separate environmental C^2 structures, there were no clear lines of authority. At that point in time in some quarters, this problem was compounded by strong resistance to the joint-staff system.

It is ironic that many of the lessons learned about joint CF command and control from the era of the 1991 Gulf War and the Oka Crisis had to be relearned later. There were some attempts to capture the lessons from that era and from Operation Saguenay, but they were hampered by the fact that, for lessons above the tactical level, the lessons-learned capabilities of both the CF and the DCDS Group were in an embryonic stage, and post-operations reports were about all that were generated after the Saguenay flood response. The lack of a proper lessons-learned process concerned people at the time, and that in itself was another lesson learned: the importance of having a cell that could actually capture these lessons.

Operation Saguenay was Henault's introduction to the DCDS Group, and it prepared him well for his job as COS J3. The next crisis to come along for him was in Zaire in November 1996. At the time, many in the international community believed there was the potential for large numbers of people who had fled the 1994 civil war in Rwanda to die in refugee camps in Zaire. Henault was involved as COS J3 in the planning of the CF's response to that perceived crisis, Operation Assurance.[14] As noted earlier, the joint-staff system was embryonic and evolving, but the decision to involve Canada in this operation was made on the spur of the moment and necessitated a quick reaction by the CF. At the beginning of November, contingency plans were started for Canada to contribute to any potential UN mission.

Given the state of the joint staff at the time, the Environmental commanders, who at this stage were still located in St Hubert (Army), Winnipeg (Air Force) and Halifax (Navy), played a major role in planning the CF's possible response in this crisis. In their assessment of what was achievable, only a small Canadian contribution could be made, for example, providing a C130 transport aircraft detachment and some command and control capability. At first it was assumed that the CF contribution would be minor, but things turned out quite differently than had been expected. On 9 November 1996, the UN Security Council voted to establish a multi-national force (MNF) to deal with the crisis in Zaire, and on 12 November Canada announced that it was prepared to take the lead, when the two most likely candidates to lead the force, the United States and France, declined to do so.

Lieutenant-General Maurice Baril, then Commander of the Army, was appointed as the MNF commander and immediately dispatched to take up his duties and consult with Raymond Chrétien, Canadian Ambassador to the United States (and ambassador to countries in this part of Africa from 1978 to 1981), who had been appointed by the Secretary-General as the UN Special Envoy to the African Great Lakes Region. Suddenly,

Canada went from being a minor MNF contributor to being the lead nation. The DCDS Group, therefore, became very engaged because of the necessity to co-ordinate the contributions of some ten or twenty nations to the mission, which was planned to consist of some 10,000 troops. A whole mechanism was created by the DCDS Group, again without the full joint-staff structure that the CF has now, to actually do that. The DCDS, then Lieutenant-General Armand Roy, went off to Stuttgart, where General Baril had set up a headquarters cell, to create a proper command and control structure and to co-ordinate the international contributions to the mission.

It was planned that the core of General Baril's headquarters would be the Army's 1st Canadian Division Headquarters, which had been made responsible for providing a deployable joint headquarters for the CF.[15] The division headquarters was selected because it had the control, communications and intelligence capabilities that were required by a lead nation to mount an operation of this size. The DCDS Group worked closely with the division headquarters and other government departments to get support for the operation. The Group was also trying to support Lieutenant-General Roy in Stuttgart by co-ordinating what he required in terms of setting up General Baril's force. In addition, the DCDS Group was trying to organize the CF command and control capability, including an airlift control element for air transport forces and a joint staff-based head-quarters structure. This latter point became controversial because it was not universally agreed upon, in the Land force or even in NDHQ, that the joint-staff structure was the right one to adopt. In the end, an ad-hoc command structure was cobbled together, but it was not based on the joint-staff structure as we now know it or even on the joint-staff system. Perhaps fortunately, given the rudimentary state of Canada's joint and combined C^2 capabilities, Operation Assurance was stillborn.

In the middle of all these activities and just as the CF was about to launch the first aircraft over to Zaire, Lieutenant-General Roy told Henault that he was going on leave for an indeterminate period. Shortly thereafter, since Roy never returned from leave and because there were no major-generals in the DCDS Group at the time, Henault, still at the rank of brigadier-general, was appointed Acting DCDS indefinitely. He was promoted Major-General early in 1997 and remained as Acting DCDS until the fall of 1997.

As Operation Assurance was winding down at the end of 1996 and Henault's thoughts were turning to improving the CF's joint and combined C^2 capabilities during a period of relative calm, he faced a new challenge: mounting Operation Assistance. In the spring of 1997, with serious flooding in southern Manitoba that threatened Winnipeg, the provincial capital, the CF was called upon to aid civil authorities in an operation co-ordinated by the DCDS Group: Operation Assistance (April–May 1997). Official DND sources describe Operation Assistance this way:

> In Manitoba, the wet spring of 1997 followed on the heels of a winter of heavy snow, and the Red River began flooding in early April. By April 20, when the provincial government requested military aid, the area surrounding the city of Winnipeg was mostly inundated and the people of the Red River Valley were beginning to lose their battle with the rising water.
>
> On April 21, the CF launched Operation Assistance, under which more than 8,500 Regular and Reserve soldiers, sailors and Air Force personnel from across Canada were mobilized to work under the direction of Emergency Preparedness Canada, helping provincial and municipal authorities and thousands of volunteers. Their primary tasks were to fill sandbags and use them to build floodwalls and breakwaters; when the water began to seep in, the next challenge was to set up and tend pumps. On May 1, the flood crested, spilling over the Winnipeg floodway and spreading through the city.
>
> By May 12, the worst was over and the troops began to withdraw; the 3rd Battalion, Princess Patricia's Canadian Light Infantry returned to Edmonton and the 1st Battalion, The Royal Canadian Regiment went back to Petawawa. On May 13, as 135 CF vehicles rolled through downtown Winnipeg on their way out of town, the citizens lined the streets, clapping and cheering.[16]

During Operation Assistance the DCDS Group used the joint-staff structure more rigorously than ever before. The key to this structure was the fielding of a joint-force headquarters, based on the Army's 1st Canadian Division Headquarters in Kingston, which deployed to

Winnipeg under the command of Major-General Bruce Jeffrey. The co-operation among the Air Component Commander, Maritime Component Commander and the two brigade commanders virtually eliminated the friction that still existed at the Environmental head-quarters' levels. Jeffery worked directly for and kept in constant contact with the DCDS, effectively bypassing the Environmental headquarters. Despite some co-ordinating problems, the operation was a success. When the Canadian public saw that the CF was able to respond in large numbers and in a joint organization to a domestic natural disaster, this acted, in Henault's view, as a real catalyst for the upsurge in public support for the CF in the post-Somalia era.

Even though the DCDS Group was busy with two major operations in late 1996 and early 1997, it was still able to undertake some significant restructuring activities. For example, changes were made to reinforce and increase the capability of the military police to do their job. Nevertheless, in this time period, the DCDS Group's joint command and control capability was not as developed as it eventually became. In retrospect, Henault believes that one of the keys to the success of the planning for operations at this time was that the Army commander (at the time Lieutenant-General Maurice Baril) and his headquarters, which had moved from St Hubert to NDHQ in the fall of 1996, were in Ottawa and able to assist the DCDS Group's response to Operation Assurance and Operation Assistance.

After Operation Assistance, the importance of the CF having a deployable joint headquarters was clear to Henault. During that operation, the Army's 1st Canadian Division Headquarters was used as a deployed headquarters because it had the command and control and intelligence co-ordinating capabilities that were required to do the job. The lessons from Operation Assistance reinforced the need for a very well articulated joint-staff system based on a deployable headquarters.

For about one year, from the fall of 1997 to the fall of 1998, Henault left the DCDS Group and worked in the air staff as Assistant Chief of the Air Staff. Even though the Army had moved its headquarters to NDHQ in the fall of 1996, it was not until the fall of 1997 that the Air Force moved its headquarters from Winnipeg to Ottawa. Henault recalled that he was able to bring to bear some of his command and control lessons from his time in the DCDS Group, especially the need for a good joint staff, on the organization of the new Air Force headquarters and on other organizational changes in NDHQ. While he was working on the air staff, the DCDS Group co-ordinated the CF's response (Operation Recuperation) to the aftermath of the ice storm that had hit parts of Ontario and Quebec at the beginning of 1998.[17]

Not long after being promoted Lieutenant-General, Henault was appointed DCDS and right away he had to oversee the CF response to the Swiss Air 111 disaster not far from Halifax in September 1998 (Operation Persistence). Official DND sources describe Operation Persistence as follows:

> On September 2, 1998, Swissair flight 111 from New York to Geneva crashed into the chilly waters of the Atlantic Ocean just off Peggy's Cove, Nova Scotia, killing everyone aboard. For 12 days, under direction from the Transportation Safety Board, Canadian Forces personnel toiled with members of the Canadian Coast Guard, the RCMP, municipal and provincial police forces, and hundreds of volunteers to recover the bodies of the 229 passengers and crew, and as much of the aircraft as could be found. While divers and boat crews searched the water and the sea bed, and teams on foot combed the beaches, the families of the deceased waited for their loved ones to be found and the world media flocked to the scene.[18]

Operations Persistence, Recuperation and Assistance demonstrated that there would be an ongoing need for the CF to conduct large-scale contingency domestic operations. Therefore, the DCDS Group initiated a series of regular domestic operations conferences to improve the ability of the three Environments —Army, Navy and Air Force — to be able to respond regionally to calls for assistance in Canada. Henault saw the ice storm (Operation Recuperation) as a defining moment for the CF in domestic operations because it was the first domestic operation to use the joint-staff structure to co-ordinate the activities of almost 16,000 service personnel, both regular force and reserve, from all three Environments in the field.

In some ways these domestic operations in the late 1990s were a precursor to the new model of "integrated" operations, involving not just DND but many other federal, provincial and municipal agencies and non-government agencies. For example, in Operation Recuperation the CF provided the support that was needed and requested by non-DND agencies. The type of support given in this operation, in trying to re-establish electrical power and get the infrastructure back up to a working level, is typical of the type of CF support envisioned for a post–9/11 domestic operation.

Besides the organizational impact of these domestic operations on the CF, there was also a leadership and command impact, which to this day continues to affect the CF, including the CDS's recent transformation initiatives. For example, during Operation Assistance, General Hillier, now CDS, was a brigadier-general and a brigade commander, and the officer, Vice-Admiral J.Y. Forcier, whom he appointed to command the first of Canada's unified commands, Canada Command, was the Maritime Component Commander in that operation. Henault believes that the continuity of experience and the continuity of command were very important factors in reshaping Canada's joint C^2 structure because those commanders had learned many lessons that came out of the operations. Even though the lessons learned were not written down and had not been catalogued for all to consume, the CF nonetheless had the experience that was well established in the minds of many of its senior leaders. They knew what had to be done to make the next operation work better, and the joint staff was able to respond again in a much more refined fashion than had ever been seen previously.

Henault was thankful that the CF had recent domestic operations experiences to lean back on and to help propel them forward, because no sooner had they finished Operation Recuperation than the government decided to contribute CF-18 fighter aircraft to the NATO air campaign in the Balkans (Operation Echo to the CF, and Operation Allied Force in NATO parlance). Official DND sources describe Operation Echo as follows:

> Operation Echo began in June 1998 when Canada sent six CF-18 Hornet fighter aircraft and approximately 125 Air Force personnel to Aviano, Italy, to help enforce a no-fly zone in the

Balkan region in support of the NATO Stabilization Force (SFOR) and Kosovo Force (KFOR), and to prepare for the 79-day NATO air campaign over the Federal Republic of Yugoslavia, which took place between March and June 1999.... Operation Echo peaked at more than 300 CF personnel and eighteen CF-18 Hornet fighters, most of them from 3 Wing Bagotville, Quebec, and 4 Wing Cold Lake, Alberta.

When Operation Echo ended on December 21, 2000, the strength of the Canadian contingent stood at about 120 Air Force personnel and six CF-18 Hornet fighters, and the return to Canada began in January 2001.[19]

During Operation Echo, the largest Canadian Air Force combat operation since the 1991 Persian Gulf War, Canadian CF-18s flew 678 combat sorties over 2,600 flying hours without loss to participating Canadian aircrew and aircraft. Despite enormous challenges, Canadian ground support staff maintained an astonishing 99.4 percent aircraft availability rate during the 79-day period of operations. The proficiency of the aircrew was also recognized by their colleagues as Canadians led over one half of the packages they flew — a proportion second only to the US Air Force.[20]

During Operation Echo, the effective use of the joint-staff system by the CF became critical to the success of Canada's contribution. It soon became clear to the CF that all elements of DND, both military and civilian, were going to be needed to support this endeavour. The joint-staff structure was crucial to supporting Operation Echo, especially in ensuring that aircraft spares and munitions and replacement pilots and technicians were available when required. In Henault's view, the CF and particularly the joint staff in its planning and co-ordination role were faced with a monumental task in Operation Echo, and they completed it successfully.

Yet the success of the CF joint-staff model was built on experience in earlier domestic operations in the late 1990s, and this experience enabled the model to evolve to a point where it could support a major combat air operation far from Canada. For example, Henault recalled an incident where the wing commander of Bagotville was having great problems re-supplying his deployed aircraft with munitions and spares during Operation Echo. At first he was going directly to the supply depot in Montreal for spares, which were then getting lost in the system because a number of organizations became involved in sending the spares over to the deployed forces without the benefit of co-ordination from the joint staff. Once the various organizations were made aware of the role of the NDHQ joint staff in co-ordinating such requests, many of the problems were overcome and most of the requirements for logistics support were satisfied.

Another important role of the DCDS, besides overseeing joint-staff activities, was providing the public face to Operation Echo. Henault was asked by the CDS of the day, General Baril, to provide the public commentary at the beginning of the operation, or to at least lay out for people what the CF were doing during the campaign. That commentary, by Henault or his COS J3, soon became an almost daily occurrence because there was a real cry from the public and the media for information on what the CF was going to be doing, especially given the significance of the operation and the fact that Canadian aircraft were going to be dropping bombs in what appeared, to some, to be a humanitarian mission.

The lesson of the requirement to provide daily media briefings during major operations was learned during the 1991 Persian Gulf War. These briefings not only met the public's need for information but also were a way of maintaining public support for what the CF was doing. At first, for Operation Echo, it was thought that the briefings would be needed for a week or two; it was never anticipated that there would be a daily briefing for a period of 79 days. And it was a lesson in itself that during major operations the DCDS had to be personally available to conduct daily media briefings on top of all his other duties. There was also an operational security dimension to the briefings, which sometimes frustrated the media because they wanted as much information as possible; however, they generally understood that there were limits to what could be provided to them.

Another lesson from Operation Echo for a major international operation was the importance of co-ordinating media briefings with the Department of Foreign Affairs so that both the military and the political sides of the operation could be presented together at the same time and place. Achieving this outcome presented some interesting challenges, however. There is a six-hour time difference between Ottawa and NATO head-quarters in Brussels, and NATO headquarters was giving a daily briefing, Canadian officials were giving a daily briefing, and officials in Washington were giving a daily briefing. Therefore, DND had to harmonize its briefings with the others to make sure that they contained related information and were therefore topical and relevant. This harmonization required a great deal of co-ordination because, at the beginning of Operation Echo, NATO headquarters gave its briefing first, Ottawa officials gave their briefing about six hours later, and then Washington officials gave their briefing about an hour after the Canadian briefing. At one point both Washington and Ottawa were giving their media briefings at the same time, which caused problems because the interested public wanted to be able to see all of the briefings. Through consultation it was agreed to separate the briefings by a few hours, and this step improved the public information situation substantially.

The joint-staff system allowed the CF to achieve its public information goals during Operation Echo because the NDHQ public affairs staff provided a member to the joint staff's J4 Public Affairs. This person not only provided the necessary support for the daily media briefings but also kept the joint staff apprised of the public information requirements for the operation.

While the NATO air campaign was unfolding, the CF was preparing for the peace support mission that was to go into Kosovo following the air campaign. Without an effective joint staff of efficient personnel with experience from previous operations, the CF would not have been able to manage all of these demands on its resources. Another important factor contributing to the success of the Kosovo operation was the availability of the three Environmental commanders and their staffs, who by that time had moved to Ottawa and were personally on hand for the many activities required to co-ordinate the operation. Henault asserted that the Kosovo operation and subsequent operations validated the concept of moving the three Environmental commanders to Ottawa. He argued that they should remain in Ottawa and that efforts to move them out of NDHQ should be resisted because it had been consistently demonstrated over the past five or six years that their presence and the presence of their staffs were vital to the effective functioning of the NDHQ joint staff. Even with modern communications systems, their physical presence in Ottawa was critical for the effective functioning of the matrix-based, joint-staff system used at that time in NDHQ, where the commanders' staffs provide essential expertise to the various components of the joint staff.

Another lesson learned from the operations conducted in 1998–1999 was that while the CF's capability to conduct operations was reasonably effective, the support capability (especially the C^2 of that capability) required to mount those operations needed more resources. This led to the formation of the Joint Operations Group (in June 2000) to provide improved command, control, communications and intelligence capabilities, and of the Joint Support Group (stood up in June 2003) to provide improved C^2 of support capabilities. The value of these new organizations and improved joint command and control processes in the DCDS Group was validated after 9/11, according to Henault, because without them the CF could not have responded as well as it did to the "war on terror."

As the twentieth century drew to a close, the DCDS Group continued to learn lessons from ongoing operations to improve the Canadian Forces' C^2 capabilities. Since the early 1990s, the CF has committed a large number of forces in the Balkans under the auspices of first the United Nations, then NATO. By the year 2000, there was a battle group in Bosnia (Operation Palladium) and a battle group in Kosovo, and air force assets (fighters, transport aircraft, tactical aviation helicopters and maritime helicopters) conducting operations and supporting land and maritime operations in the Balkans and elsewhere. At the peak of operations in the last year of the twentieth century, the CF had something like 4,500 people outside the country. While these out-of-country operations were underway, the CF was also making preparations for the anticipated computer-related problems associated with the Year 2000 (Y2K). The CF's participation in Canada's Y2K preparations, called Operation Abacus (March 1998 to February 2000), was under the direction of the commander of the 1st Canadian Division who was acting as a joint task force commander. Henault saw Operation Abacus as another key influence on the evolution of CF joint C^2 structures because it was based on the joint headquarters construct and it involved the entire Canadian Forces.

Operation Abacus was also the first time that most government departments co-ordinated their activities on a large scale in preparation for an anticipated domestic crisis. For example, the Y2K activities of organizations like Emergency Preparedness Canada, Transport Canada, Agriculture Canada, the Solicitor General and the RCMP, as well as the CF, were all co-ordinated by the Privy Council Office. Henault was involved in as many as three meetings per week with various agencies during Operation Abacus to provide the CF's input to the planning of Canada's Y2K preparations. Initially, the involvement of other government departments in Y2K planning was perhaps less enthusiastic than that of DND, but only because they had not co-ordinated their activities closely with other government departments or worked with DND much in the past. However, once everyone started to work together, they became more aware of the importance of having well thought-out plans in place and being able to respond to the anticipated crisis; with the experience of the 1998 ice storm still fresh in everyone's minds, which had affected three million people, they also knew how Y2K could affect the entire country. These Y2K preparations were the precursors to what are being called integrated operations in today's CF transformation initiatives.

Shortly after Operation Abacus, the CF recognized that its high operational tempo was overstretching its resources and that this overstretch was starting to take a tremendous toll, especially on personnel who had to re-deploy without adequate rest between deployments. At that point it was decided to consolidate CF operations in the Balkans, pull out of Kosovo and concentrate on the Bosnia area of operations. The number of personnel deployed out of country was, therefore, more than halved from around 4,500 in 1999 to less than 2,000 by the summer of 2001. These reductions were based on joint-staff assessments of what the CF was capable of doing, using information from the Environmental commanders and the support components as well as the communications and information management communities. The unanimous opinion was that in order to be able to respond to some unforeseen event in the future, and history has shown that there will be such an unforeseen event, the CF needed to reduce its commitments to sustainable levels.

General Henault's Observations

During Henault's term in the DCDS Group, as we have seen, the CF joint C^2 framework evolved from a largely ad-hoc arrangement into a more systematic and effective joint C^2 organization. A key change in this evolution was clarification of the roles of force generation versus force employment, which had been so contentious at the beginning of Henault's time in the DCDS Group. It was decided that force generation would become primarily the domain of the Environmental commanders, even though they did have some important force-employment functions, and that force employment would come within the purview of the DCDS, acting on behalf of the CDS who ultimately commands troops in the field. With the DCDS becoming the person most responsible for force employment in the CF, the principle of unity of command was respected in an unprecedented way.

This change made a tremendous difference in the CF's ability to conduct effective operations at the end of the twentieth century and the beginning of the twenty-first century, since a clearer command structure meant that the Environmental commanders and support organizations recognized they had to talk to the DCDS and place their forces under his command when required. They accepted this new situation because they understood that through the joint-staff system their expertise would be used and their influence recognized and that operations would be co-ordinated in the most effective way. Furthermore, by focusing on force generation and not concerning themselves with major force-employment responsibilities, the Environmental commanders became more effective force generators, and their authority more closely matched their responsibility.

Certain other developments also contributed to the evolution of the CF joint C^2 framework during Henault's time with the DCDS Group. For example, procedures to establish standing rules of engagement were developed, based on a set of laws of war and laws of armed combat that was revised to meet the needs of the new world order. These initiatives were part of efforts to develop a better advisory capacity for the joint staff from the legal, medical and support communities.

Another initiative undertaken at this time by the joint staff was the development of a much better system of operations orders. All types of orders, whether operations orders, deployment orders or administrative orders, were able to be handled more quickly and effectively using the joint-staff system. In the first two years of Henault's time in the DCDS Group, the joint staff was not able to track effectively CF capability; for example, the joint staff did not always know the status and location of CF strategic airlift assets or naval assets. Therefore, the DCDS Group could not respond immediately to requests for assistance from the government, whether it be evacuating Canadians from some hot spot or providing relief during some disaster. In the CF assistance after the earthquakes in Turkey, for example, the DCDS Group had to depend on the Air Command Headquarters in Winnipeg for information on aircraft availability. There was no CF-wide information system at the time, and this lack of timely information created delays that were very frustrating to those who were planning operations.

In Henault's view, the changes made to the CF's joint command and control structure from 1996 through 2001 were extremely important and form the basis for much of the way in which the CF conducts joint operations today. One particularly evident success of these changes is the credibility that they have created among politicians and their staffs. Those involved in government began to understand that the changes to the CF's joint command and control structure made the CF more responsive to political direction at a time, from 1997 through 2001, when the government increasingly wanted to be involved in international operations. Whether it was for operations based on the UN, NATO or coalitions of the willing, the CF demonstrated a capability to generate, deploy and command forces that were responsive to government direction. This capability built government confidence, right up to the highest levels, in the Canadian Forces actually being an instrument of its international political will. Such confidence in the CF was particularly evident in the Canadian response to 9/11, which had a significant military component. None of this would have been possible without the changes that were made to the CF's joint command and control structure based on the experience of 1996–2000.

Another significant change in CF's joint command and control structure was the closer integration of intelligence with operations, a relationship that did not exist in the mid- to late 1990s. Today politicians demand almost instantaneous answers to their questions about CF operations, and in Henault's experience as the Chief of Defence Staff, using the current joint-staff system he was able to provide those answers in a timely manner. Just as important, the joint-staff system allows the CF to respond quickly to a government request for support or involvement in an operation because in most cases the joint staff's intelligence capabilities and its engagement in and monitoring of policy issues and political discussions allow the CF to anticipate and plan ahead for many operations. Today, based on the joint staff's maintaining an accurate and up-to-date intelligence picture of what is going on in the world, the CF is able to respond quickly to government requests for action.

Henault concluded by stating that he believed the CF was among the best in the world at responding to taskings and providing value added to the missions in which they participated — anything from humanitarian assistance to combat. Canadian service personnel are very highly regarded in international circles for this excellence, which was built up piece by piece at the end of the twentieth century and at the beginning of the twenty-first century. Henault is convinced that the government's recent commitment of almost thirteen billion dollars over five years to the CF reflects the high level of confidence that the Canadian people and government have in the Canadian Forces. It has been a long and hard road back from the lack of confidence that the public had in the CF in the wake of the Somalia incident to the confidence that it has in the CF today. The CF's ability to generate, deploy and command forces post-Somalia has helped to recreate the credibility, the trust and the bond with Canadians that had been broken during the Somalia affair. Actions speak louder than words in gaining public confidence, and the CF demonstrated during Henault's time in the DCDS Group and as CDS that despite the challenges of limited funding and resources as well as a lack of recapitalization, it could perform its missions effectively.

Conclusions

The last decade of the twentieth century marked the transition to a new world order when the Canadian Forces began to conduct a whole new range of operations in addition those they had previously conducted during the Cold War. The CF's capability to conduct these operations evolved during that decade and was put to the test a number of times, culminating in the CF's response to 9/11 and the global war on terror. The

CF's response to the challenges of mounting operations in this era was due in no small measure to its effective joint and combined C^2 structure.

For a span of eight years, between 1996 and 2004, General Henault served in a number of key positions in NDHQ, including de facto DCDS and DCDS for four years and CDS for just over three years. During those eight years he had an important influence on the evolution of CF joint and combined C^2, and it could be argued that at the beginning of the new world order he was one of the most experienced senior practitioners of operational art in Canada. Therefore, his experiences and views are vital to understanding the Canadian joint command and control structure at that time and to understanding how they might evolve in the future. This chapter focused on Henault's time in the DCDS Group and gives his impressions of this evolution.

When Henault started working in the DCDS Group, there was a limited joint-staff capability, and the command and control of many operations was based on an ad-hoc system that involved the Environmental commanders and a very rudimentary joint staff. None of the joint-staff co-ordination mechanisms, like the Crisis Action Team that existed until recently, were in place at that time. There was also no effective co-ordination mechanism for the various CF support capabilities.

The first major operation that Henault was involved in during his time in the DCDS Group, Operation Saguenay in the summer of 1996, made him realize that the method of co-ordinating the activities of the Army, Navy and Air Force through the use of desk officers in NDHQ was not an effective C^2 framework for joint operations: there were no clearly defined lines of command, authority and responsibility. By the end of 1996 it was clear to Henault that a joint staff-based headquarters structure was required for the DCDS Group to effectively co-ordinate CF operations, but it took some time to implement this structure because of a lack of agreement among a number of stakeholders in the CF. By the spring of 1997, during Operation Assistance, the DCDS Group started to use the joint-staff structure more rigorously to co-ordinate the activities of the Army, Navy and Air Force; however, the CF depended on the Army's 1st Canadian Division Headquarters to act as a deployed head-quarters.

The joint-staff structure was modified again after Operation Assistance based on previous experience, and Operation Recuperation in early 1998 (involving almost 16,000 service personnel from all three Environments) marked further progress in the evolution of the CF's joint C^2 capability. The domestic operations in the late 1990s were the forerunners of the new model of integrated operations that is part of the CF's transformation initiatives today. They involved DND and other federal, provincial and municipal agencies as well as non-government agencies learning to work together closely. These operations culminated in Operation Abacus, which ushered in a new century and a new era of co-operation where Canadian government and non-government agencies for the first time co-ordinated their activities to an unprecedented degree to prepare for an anticipated domestic crisis. By the time of Operation Echo (June–December 2000), the CF contribution to the Kosovo campaign, the CF joint-staff system showed that it could handle combat operations as well as domestic disaster-relief operations. The CF joint C^2 capability was further enhanced with the formation of the Joint Operations Group in June 2000 and the Joint Support Group in June 2003.

The successes engendered by the constantly evolving CF joint C^2 capability were important in improving the CF's post-Somalia public image and in earning public support and trust. While normally working as DCDS behind the scenes, during Operation Echo when he was the public face of the CF, Henault made a personal contribution to the re-building of the CF's image. The briefings he and others gave during Operation Echo were effective, in part, because they were based on public affairs experience from previous operations and they featured co-ordination with other government departments in Canada and with coalition partners abroad.

The closer integration of intelligence with operations was another significant change in CF's joint command and control structure made during Henault's time in NDHQ. By using joint-staff resources to constantly monitor the world and domestic situations, the CF was and is often able to anticipate government requests for action, thereby being prepared with options for the government's selection. This better intelligence capability also allows the CF to reply quickly to government requests for information.

The combination of an improved public image and enhanced C^2 capabilities was an important factor in increasing the credibility of the CF in the minds of politicians and government officials in the post-Somalia era. This increased credibility and the CF's improved responsiveness to government direction gave the government the confidence to use the CF as an instrument of policy when appropriate.

Despite the success, based on past experience, of many operations in the late 1990s and early 2000s, throughout Henault's time in NDHQ the lack of a proper lessons-learned process concerned him. The inability to transmit lessons learned from the various operations in a systematic and coherent manner meant that to function effectively the CF's C^2 framework depended on experienced staff who could provide continuity over time based on their personal experience. This way of passing on experience, while having some value, is not as robust or as reliable in the long term as a proper lessons-learned system that provides written lessons learned, embedded in the CF training and education system.

During Henault's time in NDHQ, the CF joint C^2 framework evolved from a largely ad-hoc arrangement into a more systematic and effective joint C^2 structure. Besides the changes described above, the functions of force generation and force employment were clarified: force generation became primarily the domain of the Environmental commanders and force employment became within the purview of the DCDS. Nevertheless, Henault believes that it is vital for the commanders of the Army, Navy and Air Force to be physical located in NDHQ so that they can personally consult with other DND leaders in times of high operational tempo or crisis.

The CF has performed exceptionally well in the new world order, responding in a timely and effective manner to operations that could not have been foreseen in the Cold War era. The CF's exceptional performance was the result of many factors, especially its superbly trained personnel. However, without an effective C^2 structure to co-ordinate, command and control the CF's operations in the new world order, the CF would have been hard pressed to do as well as it did. As we have seen, the CF's C^2 structure in place today is a product of a relatively short period of evolution based on the experience of ongoing operations. This evolution was carried out in lean times for the CF, as budget cuts in previous years had reduced CF capabilities and no budget increases were planned to fund the many new operations that the government committed the CF to undertake in the post–Cold War world.

In another interview General Henault described his CDS experience of trying to transform the CF while it was making a substantial effort in the campaign against terror like "changing the tires on a moving car."[21] This might also be an apt metaphor for those years, described in this chapter, when the CF was making substantial changes to its C^2 structure at the same time as conducting operations at a tempo unprecedented for the CF in peacetime. General Henault was instrumental in guiding and influencing these changes while overseeing many of the CF's major operations at the beginning of the new world order, and his experiences will provide valuable lessons for those who follow in his footsteps.

NOTES

1   This chapter is based on the transcript of an interview conducted by Brigadier-General (Retired) Joe Sharpe with General R.R. Henault on 11 April 2005. The transcript was edited into its current form by Allan English.

2   Before unification Canada had three separate services: the Royal Canadian Navy, and the Royal Canadian Air Force, and the Canadian Army. After 1 February 1968, when the Canadian Forces Reorganization Act took effect, all Canada's armed forces were unified into a single service – the CF. While the RCN, the RCAF, and the Canadian Army no longer existed as legal entities, people often referred to the navy, air force and army in everyday usage. However, to emphasize the point that Canada no longer had three services, Department of National Defence bureaucrats coined the rather awkward term "*environment*," based on the environments in which the sea, air, and land components of the CF operate, to describe these three components of the CF. Since there is only one military service in Canada today, the CF, official DND publications sometimes use the noun "*environment*" and the adjective "*environmental*" when referring to the sea, air, and land components of the CF. Nonetheless, the terms *Canadian Army, Navy* and *Air Force* have now returned into official usage.

3   Canada, Department of National Defence (DND), *The Aerospace Capability Framework* (Ottawa: Director General Air Force Development, 2003), 43.

4   "About DND/CF, Budget," DND Web site, http://www.forces.gc.ca/site/about/budget_e.asp (accessed 4 May 2005).

5   DND comprises both the CF and the civilian employees that support the CF, "The National Defence Family," DND Web site, http://www.forces.gc.ca/site/about/family_e.asp (accessed 1 May 2005).

6   David Detomasi, "Re-engineering the Canadian Department of National Defence: Management and Command in the 1990s," *Defense Analysis* 12, no. 3 (1996), 329–30.

7   These issues are discussed in more detail in G.E. (Joe) Sharpe and Allan English, *Principles for Change in the Post–Cold War Command and Control in the Canadian Forces* (Kingston, ON: Canadian Forces Leadership Institute, 2002).

8   "The National Defence Family," DND Web site, http://www.forces.gc.ca/site/about/family_e.asp (accessed 4 May 2005).

9   See for example DND, *Annual Report 2002–2003* [of the Ombudsman for National Defence and the Canadian Forces], http://www.ombudsman.forces.gc.ca/reports/annual/2002-2003_e.asp#unfair (accessed 1 May 2005).

10   Figure from DND, *Canada's International Policy Statement: Defence* (19 Apr 2005) http://www.forces.gc.ca/site/reports/dps/pdf/dps_e.pdf, p. 7, (accessed 1 May 2005).

11   This part of the report on the post–Cold War CF and Air Force operations is based on Rachel Lea Heide, "Canadian Air Operations in the New World Order," in *Air Campaigns in the New World Order: Silver Dart,* Canadian Aerospace Studies Series, vol. 2, ed. Allan D. English (Winnipeg: Centre for Defence and Security Studies, 2005), 77–92.

12   Figure from Heide, "Canadian Air Operations in the New World Order," 91.

13   Figure from Heide, "Canadian Air Operations in the New World Order," 92.

14   For more details on this operation see Michael A. Hennessy, "Operation Assurance: Planning a Multi-national Force for Rwanda-Zaire," *Canadian Military Journal* 2, no. 1 (Spring 2001), 11–20.

15   See C.J. Weicker's paper herein for more details on the evolution of the CF's joint deployable headquarters capability.

16   DND, "Operation Assistance," http://www.forces.gc.ca/site/operations/assistance_e.asp.

17   For details of this operation, see DND, "Operation Recuperation," http://www.forces.gc.ca/site/operations/recuperation_e.asp.

18   DND, "Operation Persistence," http://www.forces.gc.ca/site/operations/persistance_e.asp.

19   DND, "Operation Echo," http://www.forces.gc.ca/site/operations/echo_e.asp.

20   David Bashow et al., "Mission Ready: Canada's Role in the Kosovo Air Campaign," *Canadian Military Journal* 1, no. 1 (Spring 2000), 55, 58–9.

21   Allan English and Joe Sharpe, "Lessons Learned from the Perspective of a Chief of the Defence Staff," *Bravo Defence*, 13.

CHAPTER 7

COMMAND AND CONTROL CANADIAN STYLE: THE NEW MEDIUM-POWER DILEMMA

Commander Kenneth P. Hansen

In 1986, Rear-Admiral J.R. Hill described the "medium-power dilemma" in his landmark book, *Maritime Strategy for Medium Powers*. Unlike most maritime strategy, which is written from the perspective of maritime superpower nations that place high value on naval forces, Admiral Hill's work clearly set out the hard choices that confront "lesser powers" and those states with more ambivalent attitudes towards the major investments of time, labour and capital that sea power represents. He described the difficult and sometimes perilous balancing act that occurs when the demand for a national naval capability conflicts with politically negotiated alliance security arrangements.[1] Typically, domestic sovereignty tasks do not call for the types of high-end combat capability and endurance that "blue water" alliance naval tasks require. The history of political debate on naval procurement for Canada includes many raucous exchanges between the proponents of a centralized strategy, who looked mainly to alliance demands (or in times past, Imperial "obligations") to set the character and capability of the navy, and "nationalists" who sought a decentralized or domestically oriented strategy and, by doing so, hoped to attain a degree of institutional control that would avoid "foreign entanglements." The same arguments have also been made for and against all other types of Canadian military procurement plans.

The debate over the question "What capabilities shall Canada's Navy have?" has been typical of Hill's dilemma. The result, a product of our consensus-driven democracy, has been a hybrid force structure that tries to meet both domestic and foreign requirements in some limited and cost-effective way. Historically, Canada's commitments to coalition operations have been composed of specialist forces — units and small formations that are quite capable in one or more warfare areas but have little or no capability in others. The home defence requirement was also covered by the same forces, which were often deployed elsewhere, as the lack of a perceived threat to Canada allowed the luxury of dual tasking. In the relatively stable geo-political environment that ended with the close of the Cold War, such specialist forces were capable of meeting the fairly predictable tasks that arose with great regularity. The limited utility of Canada's niche military capabilities also helped to limit their employment options, which had significant political advantages in avoiding risk, casualties and escalation.

Even more problematic, however, has been the question "What are Canadians prepared to pay for in the way of military forces?" Professor John Treddenick has argued that the budgetary process has actually been the main driving force behind decisions on defence policy.[2] As the cost of sea power continued to escalate[3] and the perceived benefits from naval defence construction programs dwindled,[4] Canada's naval force structure shrank steadily. Virtually every naval program has been reduced from its original projections, all have been delivered late, and a notorious few have been cancelled outright. Canada's dilemma during the Cold War was how to balance domestic defence requirements versus alliance commitments on a meagre budget.

The reduced defence budget has forced a serial retrenchment by the Canadian military to preserve what it perceives as its institutional core capabilities, which are, not surprisingly, centred on its traditional specialist roles. One consequence has been that the term *multi-purpose forces* has supplanted *general-purposes forces* in Canadian defence documents. This is a subtle doctrinal indication that some military capabilities and operations are permanently outside of our "normal" range of expertise. Such reductions in force structure, capability and doctrine could not help but influence the command and control capabilities of the military they were meant to direct. A special target of the budget-reduction process has been both the numbers of headquarters and the size of the staffs in those that remained after the reduction process had been completed. The object was to enhance the "tooth to tail ratio" by making reductions in the "bloated" headquarters organizations.

The end of the Cold War has brought a new and challenging security environment that lacks the predictability of the old geo-strategic arrangement with which we had become so familiar and comfortable.

Most discussions about the new security environment have tended to focus on the information and technology demands embodied by the Revolution in Military Affairs (RMA).[5] The command and control arguments are most often centred on the same sort of material demands that network-centric warfare and a co-operative engagement capability (CEC) will make on Canadian headquarters organizations and command systems.[6] However, quite apart from the technological issues, there are also intellectual problems that have a seldom-considered human aspect of the RMA. There is definite evidence to indicate that the lack of national joint doctrine and inadequate staff officer development will become the basis of a new medium-power dilemma for Canada. These problems may rekindle the old centralization versus decentralization debate and could have a profound influence on Canadian defence policy and force structure. But, unlike the RMA and NCW issues, these doctrinal and staff issues have already had important Canadian historical antecedents. In fact, these two issues predate the formation of the Canadian Navy itself. But, despite their age, they are still relevant to the demands being made on the Canadian Forces in the new security environment of the post–Cold War era.

Recent analysis of Canadian involvement in a proposed UN deployment (Operation Assurance) to Rwanda and Zaire has revealed serious problems of headquarters establishment and staff preparedness to undertake the leading role in a major operation. Michael Hennessy's recent article in the *Canadian Military Journal* stated:

> The doctrine assumed that Canada would not be the lead country, and therefore we had no permanent mechanism for responding to the challenges of being one.... Both the government and the CF underestimated the requirements for the organization of an MNF...the staff for both [national and deployed-force headquarters] had to be drawn essentially from the same small pool of personnel...the planning was too chaotic to permit rational development of an international force under our leadership.[7]

Hennessy concluded that the cause of this unpreparedness was Canada's traditional role as a contributor of specialized forces to coalition operations led by other nations. This arose as a natural consequence of

Canada's old approach to its middle-power dilemma. Yet, the Minister of National Defence has stated on several occasions that the new direction for the Canadian Forces is to be prepared to act rapidly in response to a global crisis.[8] The object will be to be among "the first in and the first out," thereby avoiding the type of interminable and unsustainable (and expensive) deployments that have been typical of the past forty years.[9]

This new plan to develop a capability for rapid global deployment at short notice requires high levels of readiness and self-reliance for it to be workable. The analysis of Operation Assurance highlighted the reluctance of other UN member nations to contribute to such an undertaking and the critical shortfalls in many capability areas that resulted. A Canadian ability to deploy with assurance and at short notice will require a high degree of autonomy and self-sustainment capability because, in the prevailing uncertain circumstances, even our most reliable allies may be as reluctant as they were during the Rwanda-Zaire crisis. These characteristics of readiness and autonomy are not consistent with the history of the CF, which has been one of integrating our forces into other nations' force structures to achieve the necessary degree of combined capability. Interestingly, Hennessy's article specifically identified two critical shortfalls: doctrine development and sufficient headquarters staff to do the critical planning work.

Historically, the CF has been weak in doctrinal development. Very little original, independent, Canadian naval or joint doctrine has ever been written. The Canadian approach has been to contribute to NATO doctrine development through the usual system of working groups and committees, and to adopt British and American doctrine as the situation dictated. An approved Canadian version of *British Maritime Doctrine* or the US Navy's NWP series of manuals does not exist.[10] This doctrinal weakness was very evident during Operation Assurance. Having spent their entire careers concentrating on only certain aspects of operations planning and execution, the undermanned headquarters staffs proved to be inadequate to the larger task asked of them. The high degree of readiness required to conduct such short-notice operations absolutely requires fully manned and prepared headquarters staffs who have the necessary generalist orientation, experience and doctrinal guidance to pull off such a demanding task. The natural tendency in Canada's specialized military has been to place a very high priority on tactical proficiency. In the new force concept, much higher emphasis will have to be put on such diversifying experience as foreign exchange and liaison postings, post-graduate and staff college education, and joint employment that takes the prospective staff planning officer out of his or her normal service environment. The new criterion for advancement must be one of broadly based operational-level knowledge and experience, not superlative skill in a narrow tactical field of expertise.

One of the mechanisms used to help prepare for unexpected tasks is the pre-planning of contingency operations plans (COPs). These are plans that can be drawn upon when required as the framework of a more detailed and task-tailored Operations Order. However, their production is a very large and long-term commitment that cannot be accomplished with only a skeleton staff. Likewise, the plans must be written in terms that are comprehensible to the ultimate users, and they must be pertinent to the users' capabilities. A plan written in the United States for a marine expeditionary unit from Little Creek, Virginia, will be of little use to a Canadian battle group from Petawawa, Ontario. Only officers with a sound understanding of joint doctrine and national operations procedures are capable of producing effective and coherent COPs. As Operation Assurance demonstrated, such people are in desperately short supply.

The confusion and unpreparedness that came to light during Operation Assurance was not a solitary event. During the Gulf War, typical Canadian concentration on specialist roles led to many questionable decisions and difficulties with the command and control arrangement. An NDHQ study by the Director General of Programme Evaluation (DGPE) found "a surprisingly large variety of distinct command philosophies within our unified force."[11] The study also attributed some of this problem to the use of ad-hoc units, which had to be used for reasons of manning flexibility and economy, stating that formed units would have been preferable. The joint force headquarters (JFHQ) in Bahrain was found to be a typical example of such an ad-hoc unit.[12] A permanent, deployable JFHQ did not exist at that time, although one has recently been formed in Kingston, Ontario. The significant problems of forming a joint force commander from the disparate contributions of the various commands that had *never* operated together before in that capacity can be imagined. The exasperation of the author of the DGPE report fairly jumps from its pages in the following passage:

> Despite the promulgation of the CDS's command concept for the Gulf, a number of comprehensive briefings to senior staffs, and lengthy discussions involving key personnel, supporting commands did not all support the need for a JFC in the Gulf. Differences in environmental doctrine, terminology, and a belief that control was over-centralized under the J-staff at NDHQ led to a constant need to re-iterate the CDS's intended C^2 relationship.[13]

The CF's history of service-specific specialist roles had an obvious effect during the Gulf War in the basic lack of familiarity with joint operations doctrine, and even went so far as to create an atmosphere of distrust between the supporting commands and the nation command authority. That there would have been a similar distrust within the theatre between the component commands would have been a natural extension of the historical situation and the problem described above.

Probably the most critical problem was the choice of location for the JFHQ in Bahrain. The JFC chose it because "Bahrain presented a preferable location than Qatar, since it would facilitate communications with US NAVCENT, whose command ships, USS *Lasalle* and USS *Blue Ridge*, both operated from there."[14] Rear-Admiral Summers' bias as a naval officer was clearly to integrate his operation with that of his major naval coalition partner. There had been proponents of placing the Canadian JFHQ in Riyadh together with US CENTCOM and British Forces Middle East. Despite this, Bahrain was chosen for what is only vaguely described as "a combination of diplomatic, political and military reasons," and placing a Canadian air liaison officer of lieutenant-colonel rank in the air component headquarters at CENTCOM was deemed an adequate compromise.[15] But, as the national JFC, Summers had a higher responsibility to ensure that all the Canadian theatre-level interests, not just air-related issues, were served in Riyadh, Saudi Arabia. The US Navy's maritime component commander in the theatre, Vice-Admiral Arthur, also decided not to co-locate his headquarters with that of the theatre commander, General Schwarzkopf, and for this he has come under serious criticism. Marvin Pokrant has argued that Vice-Admiral Mauz,

Arthur's predecessor, should have located his headquarters in Riyadh and that his decision not to do so resulted in the US naval forces not being used to their full advantage during the war.[16] Pokrant also records that Admiral Arthur knew before he arrived in the theatre that he should have been located in Riyadh but that the shortness of time precluded establishing the required communications arrangements. After the fact, Admiral Arthur assured General Schwarzkopf that "if he could wipe the slate clean, he would be in Riyadh."[17]

For his part, Admiral Summers established his headquarters before hostilities commenced, and Canadian naval embargo operations were a model of efficiency.[18] However, when the most critical C^2 moment came and the decision to undertake hostilities arrived, the Canadian naval forces were caught virtually unaware, and incredibly, "for a variety of reasons they had not formalised plans to join the U.S.–led offensive."[19] The lack of both a USN and a Canadian naval high command presence in Riyadh played a large part in this lack of awareness and the less-than-optimum naval performance during hostilities. As a counter-argument, Admiral Summer's chief concern had been that the Canadian naval contingent not be split up among the larger USN carrier task forces and, thereby, lose their Canadian identity, as had been the case during the Korean War. This was, however, a misinterpretation of the situation during that earlier war.

The first Canadian commander of Canadian Destroyer Squadron Far East (CANDESFE) during the Korean War, Captain Brock, argued vehemently with Rear-Admiral Andrewes, the British component commander, against the integration of the forces of the Canadian Navy with those of the Royal Navy. He was largely successful in this endeavour. However, Brock's successor, Commander Fraser-Harris, acquiesced to Admiral Andrewes' demands, partly because of his relatively inferior rank to Captain Brock and partly because of his strong personal connections with the RN. Ultimately, although the command relationship situation with the RN was rectified after command of CANDESFE was subsequently transferred to Commander Plomer and after a visit to the theatre by Canadian Rear-Admiral Creery, the real reason the force's national identity was so difficult to discern was that the "Island Campaign" in which the Canadian ships were employed for the later part of the war did not require anything more than single-ship operations.[20] There simply was not a

Canadian naval employment option that required CANDESFE's destroyers to be used as a group.

The most important issue with CANDESFE's employment during the Korean War was the lack of dedication by Commander Fraser-Harris to Canadian command-relationship doctrine and his willingness to acquiesce to British authority. These attitudes are traceable back to the earliest days of the Canadian Navy and the arguments over Imperial centralization or national decentralization.[21] When, finally, the issue was resolved for the creation of the Royal Canadian Navy, versus direct contributions to the Royal Navy, the Imperial Defence Staff contrived a system of staff training that would, without directly affecting the naval policy of the Dominions, bring about a certain unity in method and operational philosophy (doctrine) that the direct attempts at Imperial direction of Dominion forces had been unable to achieve:

> An exchange of [British and British-trained Canadian] staff officers could make the work of the General Staff in the largest sense the work of a military mind which survey the defence of the Empire as a whole...and would bring about uniformity of pattern in organization and in weapons, and in other details regarding military matters, which is to some extent essential if there is to be effective co-operation in a great war.[22]

The long history of Canadian naval staff and seamanship training at the knee of the RN carried on well past the Korean War. Commander Fraser-Harris's "loyalty" to the British way of doing things was only one small product of the RCN's cultural loyalty to the RN. Such loyalty to foreign doctrine, whether it be NATO, British or American, has not yet been totally replaced by a Canadian-developed doctrine. But if the plan for the CF is to be as the Minister of National Defence says, and a truly responsive and capable Canadian rapid-reaction capability is to be the result, national doctrine must be produced that will guide the efforts of the Canadian headquarters staff and deployed forces. Their staff training and education in the operational art must be the product of a Canadian institution responsive to a Canadian strategy that directs the efforts of the entire defence team.

The new dilemma for Canada and its military will be how to develop the body of doctrine and maintain the levels of staff readiness and preparedness to plan and conduct rapid-reaction operations on a strictly limited budget. The stated aim of limiting the number and size of headquarters and their staffs works directly against the accomplishment of the desired military goals. A healthy surplus of officers in both the permanent JFHQ and in the supporting headquarters is necessary to ensure that they can become appropriately experienced and educated to develop the required doctrinal manuals. Moreover, the surge of planning activity that has to be undertaken for short-notice missions requires a significant surplus of staff capacity over and above that which is required for the conduct of routine activities. The old and overused maxim of "doing more with less" simply will not work. Without the two essential ingredients of appropriate Canadian doctrine and experienced staff officers in adequate numbers, Canada will be forced to rely on foreign sources for doctrinal guidance and for essential military staff skills to mount and conduct operations, which will be led and controlled by other countries. Unless the demands of Canada's new medium-power dilemma can be addressed, the problems that arose during Operation Assurance are virtually certain to recur.

NOTES

1   J.R Hill, *Maritime Strategy for Medium Powers* (Annapolis: Naval Institute Press, 1986), 199–212.

2   John M. Treddenick, "The Defence Budget," in *Canada's International Security Policy*, eds. David B. Dewitt and David Leyton-Brown (Toronto: Prentice-Hall, 1995), 413–54.

3   Philip Pugh, *The Cost of Seapower: The Influence of Money on Naval Affairs from 1815 to the Present Day* (London: Conway Maritime Press, 1986), 154–70.

4   Bertrand Marotte, "$10B Frigate Project Foundered, Study Says: Grandiose Program a Dismal Failure, Federal Report Admits," *Ottawa Citizen*, 14 October 1998, http://www.ottawacitizen.com/national/981014/1934954.html.

5   Gary Garnett, "The Canadian Forces and the Revolution in Military Affairs: A Time for Change," *Canadian Military Journal* 2, no. 1 (Spring 2001), 5–10.

6   Mark Tunnicliffe, "The Revolution in Military Affairs and the Canadian Navy in the 21st Century," *Maritime Affairs* (Spring/Summer 2000), 3–7.

7   Michael A. Hennessy, "Operation Assurance: Planning a Multi-National Force for Rwanda/Zaire," *Canadian Military Journal* 2, no. 1 (Spring 2001), 11–20.

8   Robert Fife, "Military Plans Rapid-Reaction Force to Respond to Global Crises: The Way of the Future," *National Post*, 14 January 2000, 1.

9   John Geddes, "The Price of Peacekeeping: Heavy Spending in the Balkans Is Raising Difficult Questions About What Role Canada Can Afford to Play in World Affairs," *Maclean's*, 12 February 2001, 26–27.

10   The closest thing to a Canadian naval doctrine publication is the *Naval Doctrine Manual* (MCP1) produced at the Canadian Forces College, Toronto. It is not, however, distinctly Canadian in nature and is more of a survey of existing sources outside Canada.

11   DND, "A Case Study: Operation Friction (The Gulf Crisis)" 1258-99 (DGPE), NDHQ secret (declassified) report dated 13 October 1993, 41–59, contained in "Strategic Operational Management in the Gulf War: An NSSC Case Study" (Toronto: Canadian Forces College, 1999), 67.

12   Ibid., 68.

13   Ibid., 67.

14   Ibid., 44.

15   Ibid., 45.

16   Marvin Pokrant, *Desert Storm at Sea: What the Navy Really Did* (Westport: Greenwood Press, 1999), 208.

17   Ibid., 174.

18   Duncan E. Miller and Sharon Hobson, *The Persian Excursion: The Canadian Navy in the Gulf War* (Mississauga: Arthur-Jones Lithographing, 1995), 155–80.

19   Jean H. Morin and Richard H. Gimblett, *Operation Friction, 1990–1991: The Canadian Forces in the Persian Gulf* (Toronto: Dundurn Press, 1997), 144, 179–80.

20   Edward C. Meyers, *Thunder in the Morning Calm: The Royal Canadian Navy in Korea, 1950–1955* (St. Catharines: Vanwell, 1992), 124-5, 152.

21   Richard A. Preston, *Canada and "Imperial Defense": A Study of the Origins of the British Commonwealth's Defense Organization*, 1867–1919 (Toronto: University of Toronto Press, 1967), 89–119, 344–86.

22   Ibid., 360.

CHAPTER 8

COMMAND AND CONTROL DURING PEACE SUPPORT OPERATIONS: CREATING COMMON INTENT IN AFGHANISTAN[1]

Howard G. Coombs and General Rick Hillier

From Plato to NATO, the history of command in war consists essentially of an endless quest for certainty — certainty about the state and intentions of the enemy's forces; certainty about the manifold factors that together constitute the environment in which the war is fought, from the weather and the terrain to radioactivity and the presence of chemical warfare agents; and, last but definitely not least, certainty about the state, intentions, and activities of one's own forces.

Martin van Creveld

The original concept of *network-centric warfare* (NCW) articulated by Cebrowski and Garstka in 1998 seemed to be the fulfilment of the search for certainty articulated in the writings of Israeli researcher Martin van Creveld. They argued that the then ongoing shift in "platform centric" to "network centric" warfare had created a revolution in military affairs. This fundamental change in how wars are fought and won was an extension of the "co-evolution of economics, information technology, and business processes and organizations" of American society in the 1990s.[2] From these developments emerged a visualization of business and, subsequently, military organizations that were self-synchronizing, adaptable and unencumbered by hierarchical control measures. This concept is reliant on technology and requires networked information architecture with the ability to sense, process and act more quickly than competing systems. In a similar vein, the Canadian Forces is currently examining concepts of *network-enabled operations* as potentially key facets of its future force doctrine. However, while not promulgated in doctrine, the networks utilized by the Canadian Forces during peace support operations (PSOs) are not solely reliant on technology but are hybrid, consisting of a mixture of information and social networks that achieve similar effects. This can be demonstrated through an examination of the networks established by the Canadian-led International Security Assistance Force (ISAF) Rotation V in Afghanistan during 2004.[3]

The state of affairs in Afghanistan at the time of the first large-scale Canadian deployment with ISAF Rotation IV in 2003 was stable but fragile. Afghanistan had endured decades of strife and friction throughout its borders, which had destroyed any confidence in the ability of central authorities to be able to address issues of governance. At that time an American officer described the situation unambiguously:

> There are clearly things that need doing here — from basic services to CCC [Civilian Conservation Corps] — like projects (roads, reforestation, irrigation, etc.), to building of every governmental institution.... We need to disarm...and teach them about other ways of life besides poppies and illegal...check points...from which they make a living and fund and pay their armies.... But it's not about physical building (although that's part of it), it is about building the ability for Afghans to take care of themselves from public health to judicial systems. These are much harder to grow than trees or roads or schools.... [4]

The dilemma was highlighted by Canadian prime minister Lester Pearson in his 1957 Nobel Prize lecture: "Our problem, then, so easy to state, so hard to solve, is how to bring about a creative peace and security which will have a strong foundation."[5] In order to address these challenges, it is essential to produce societal renewal by understanding the nature of the culture being rebuilt and by working within that society.[6] This requires focusing on the processes that permit strengthening and development of internal structures, processes that are not predicated on technical connectivity but on the establishment and maintenance of people-to-people contacts. One can thereby endeavour to ensure balanced efforts by all agencies to create the circumstances necessary for success. The roles played by internal and external participants in the process bear particular scrutiny because local ownership must be included in these networks as they evolve and mature. Military commanders and staffs must recognize that there is a need to subordinate the martial aspects of the intervention to the imperatives of supporting the efforts of reconstruction, and provide the security that will encourage and enable these activities. While technology is important in establishing connectivity, in an environment that does not have a homogeneous technical network available to all participants, networks are established as required and are hybrid. Great efforts are required to construct these hybrid networks in order to create co-operation and interaction with a myriad of other groups and a cohesive and focused plan that fosters unity of effort. Unfortunately this unifying concept rarely exists as a cohesive whole in either national or alliance strategy. Nevertheless, as a result of the Canadian experience of PSOs since 1992, the Canadian Forces (CF) has developed the institutional knowledge to address these complex dilemmas.[7]

The Canadian vision of command and control during PSOs was developed from successful Canadian organization and leadership of the United Nations Emergency Force (UNEF) I in 1956 and succeeding "traditional" peacekeeping missions.[8] All of these missions were predicated on obtaining consensus from the belligerent parties through mediation and negotiation. Avoidance of conflict was the rule, and the maintenance of peace (or at least an absence of overt violence) was normally sufficient to achieve the objective of the mission. National and alliance concerns were primarily the containment and de-escalation of fighting, and as a post-war middle power Canada had justifiable pride in such a role within the UN and NATO.[9] In this stable environment, time-honoured hierarchical command and control mechanisms were sufficient to ensure that missions were successful.

Nonetheless, in the volatile international setting of the post-modern era it became obvious that the established Canadian vision of peacekeeping needed to change. The goal of intervention was no longer simply a cessation of violence between organized entities to avoid potential nuclear conflagration, but rather, reconstruction, renewal and development leading to functioning nation states that could act autonomously within the community of nations. Military operations must assist with creating the conditions for a durable and lasting peace in joint, multi-national and multi-agency environments, with numerous state and non-state actors involved in resolving the crisis. The military contribution to this goal is not limited to the separation of consenting belligerents, but it has become exceedingly complex. As a result, it has been acknowledged that the CF must establish connections with the Department of Foreign

Affairs and the Canadian International and Development Agency to address the modern dilemmas of post-conflict Afghanistan. Then defence minister David Pratt articulated this idea in an address to the Conference of Defence Associations Institute during 2004:

> One of the things that distinguishes Afghanistan from previous Canadian Forces missions is the unprecedented co-operation we're seeing between the Canadian Forces, the Department of Foreign Affairs and the Canadian International Development Agency. In fact, from the standpoint of future Canadian international engagements, Afghanistan is serving as a model for the government's 3D approach to international affairs — the three Ds being defence, diplomacy and development.[10]

The 3D approach has required an increased level of interoperability between departments that lack a common information infrastructure. The result has been dependence on social networks and the establishment and co-ordination of decentralized operations. The sensors, processors and systems used to create lethal and non-lethal effects in such an environment mirror many of the principles of network-centric warfare.

This phenomenon is similar to the Canadian experiences in overseas operations, such as the later phases of Canada's mission to Bosnia during Operation Palladium, and in domestic operations, such as those conducted during the Winnipeg flood or the Year 2000 (Y2K) contingency planning. The creation of these hybrid networks is normally predicated on the support of military organizations to multi-agency efforts in complex environments. Military headquarters are trained, structured and resourced to provide the necessary functions that will encourage the establishment, maintenance and co-ordination of all activity. Intertwined with these aspects is the realization that efforts to ensure a solidly constructed peace in Afghanistan require a great deal of perseverance and patience. Such determination is extremely difficult to sustain over great lengths of time, particularly with the improvised nature of the networks that are established. As a result, military commanders and planners must be prepared to provide moral and physical support to the other non-military agencies involved in the effort. Incidents that have occurred in other PSOs (such as the riots at Drvar, Bosnia, in April 1998) have posed great challenges to international and non-governmental organizations and may result in a temporary or even permanent cessation of activities unless the necessary partnerships are in place.[11]

These experiences provoke a great deal of thought on the exact nature of organizing and directing military activities that assist with peaceful societal rejuvenation. The endeavours need to become a collective effort by a variety of organizations within the context of an overarching strategic concept. The successful application of military and other aid not only requires an implicit understanding of the nature of societal reconstruction and renewal required for a particular region, but also must be linked to structures that provide focus to all entities within a given country. In the absence of these conditions the military may then have to provide the assistance and impetus necessary for the formulation of an overarching system, or networked community, with an ultimate objective of resolving "the removal of the causes of war."[12] Success will require the creation of shared vision, a common intent, and the establishment of mutual trust amongst all participants. This approach was used during ISAF V and was the product of Canadian experiences with the complex dilemmas posed by the CF's post-1992 PSOs.

Prior to the transition between ISAF IV and ISAF V in January 2004, the situation in Kabul had been stabilized through intensive patrolling.[13] The time was propitious to expand beyond tactical activities and implement a long-term plan. In a meeting preceding the NATO changeover President Hamid Karzai, the leader of the Afghanistan Transitional Authority (ATA),[14] expressed concerns regarding four principal factors that he felt would undermine the influence of ATA. First, there were internal and external threats to the fledgling Afghanistan National Army from a variety of sources. Second, the lack of human capital, that is, educated and trained people, within the ministries of the government and the security forces of ATA posed great challenges to the governance and security of the country. Third, there was a need for the promulgation of positive information supporting the efforts of ATA. Last, and most important, the absence of unified action by the multitude of governments and organizations in Afghanistan had resulted in a dissipation of development efforts and effects. As a consequence, in order to expand ISAF V beyond tactical and short-term activities, it was evidently necessary to deal with these concerns.

The most important area to initially resolve was President Karzai's concern regarding the lack of unified effort amongst those involved in the regeneration of Afghanistan. On account of this apprehension and as a result of Canadian experiences in the former Yugoslavia, it was recognized that the primary focus of ISAF V ought to be the development of a strategic objective supporting the harmonization of the international community's efforts in the reconstruction and redevelopment of Afghanistan. The initial construct was created by a Canadian-led planning team who reviewed the mandates of all the major organizations in Afghanistan and compiled a list of the objectives of each of these agencies. This proposal was first known as the Structured Process for the Harmonized Development of Afghanistan (SPDF) and later, the Investment Management Framework (IMF). Depicted in Figure 8.1, the independent but interrelated themes culminate in one national goal or "end-state" and were represented within the IMF as five strategic lines of operation, each with a clearly defined start point and final objective. Each line of operation (Security; Islamic Republic Governance; Rule of Law; Building Social and Human Capital; and National Economy and Physical Infrastructure) contained multiple objectives: short-, mid- and long-term. The goal of each line of operation would be attained gradually through these three successive phases.[15] The IMF model was offered to President Karzai and ATA in order to further assist in developing a common national vision and unity of effort through consensus building and co-ordination of all parties. It also provided a method of prioritizing the challenges facing ATA, with the purpose of eventually building a legitimate and functioning state that provided for the security and prosperity of its citizens and contributed to regional and global stability.

The IMF and derivative products that outlined the tasks and interdependencies required to achieve the objectives of the five thrust lines were briefed to ATA and various members of the international community, including the UN Assistance Mission in Afghanistan (UNAMA). There was general acceptance of the concepts and principles that underpinned the IMF. Next, members of the Canadian-guided ISAF V planning staff assisted ATA with incorporating this information into the national priority programs (NPPs). With the assistance of ISAF staff, about 40 percent of the IMF was actualized through the programs. This was the initial step in creating a shared vision.

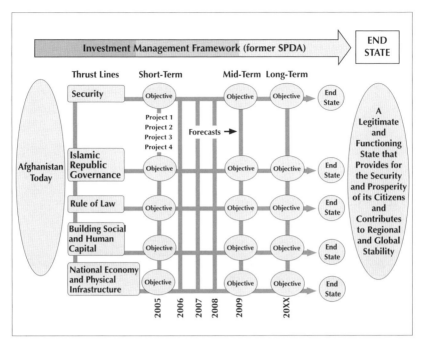

FIGURE 8.1. STRUCTURED PROCESS FOR THE HARMONIZED DEVELOPMENT OF
AFGHANISTAN AND LATER THE INVESTMENT MANAGEMENT FRAMEWORK

The national priority programs were created by ATA towards the end of
the first two years of its administration to move reconstruction efforts
from discrete and easy-to-implement projects intended to provide
immediate relief to those that would contribute to a sound future vision
for the people of Afghanistan. The overarching elements of this vision
were first outlined by President Karzai at a number of international and
domestic conferences, and he proposed that policies entailing large
amounts of funding should be carried out through a coherent program of
development rather than through the financing of discrete projects. This
idea became the National Development Framework, and its component
parts became the NPPs. The unifying principle of the NPPs was that
donor aid should be allocated through the national budget process so that
the capabilities of Afghan institutions could be consistently and
systemically increased. The national priority programs were designed to
move Afghanistan from a position of recovery and reconstruction to that
of sustainable development. As the foundation of Afghanistan's future,
they were designed to address a number of areas that promote prosperity
and deal with poverty. Additionally, the NPPs would ensure that aid

efforts would be "transparent, effective and accountable" to the populace of Afghanistan.[16]

While a great deal was accomplished with creation of the IMF and incorporation of a sizable portion of its contents within the NPPs, there was a requirement for a nationally unified concept that clearly articulated a strategy for prioritization and subsequently focused development. The logical next stage was the creation of a national concept for structured regional development. This concept would link the resources of a strategy with the methods of the NPPs to actualize the goals of the IMF. The resultant national strategy would provide crucial guidance and impetus for multi-agency co-ordination that would facilitate the construction of operational-level planning for ISAF.

As a result of discussions with the ATA Ministry of Finance, the commander of ISAF V agreed to assist Finance Minister Dr. Ashraf Ghani and his staff with the analysis of this strategic problem. Two officers deployed from Canada during June to August 2004 to augment ISAF V capabilities in fulfilling the obligation. Their work and that of other members of ISAF V resulted in the concept paper "Creating a National Economy: The Path to Security and Stability in Afghanistan,"[17] which used the existing work of ATA as a model for creating a legitimate and functioning Afghanistan. Without changing any of the established and accepted programs and policies, the paper advocated unified action of all involved agencies within an overarching security context:

> The fundamental issue right now is security...because there is clearly right now — given the security questions — reluctance [to make capital investments in the country].... Where it may not be possible to secure the entire country at one point, you could create zones of security where enhanced economic activity could be fostered.[18]

In a regional and co-ordinated manner that enabled the prioritization and allocation of resources, this approach addressed the disintegrating influences affecting the country. Furthermore, it demonstrated to the Afghan people the commitment of ATA and the international community to national reconstruction and state building.

The ideas contained in the concept paper were derived from an analysis of obstacles to success, which was conducted utilizing various official documents, but primarily the Afghanistan Transitional Authority paper "Securing Afghanistan's Future."[19] The greatest challenge to the re-establishment of Afghanistan as a functioning state was the lack of confidence by the international community and the Afghan populace in nation-building efforts. Four disintegrating influences contributed to this lack of confidence. The first was the political dissension of regional leaders who advocated local interests above that of the nation. The second was military and could be seen in the existence of the various disruptive non-governmental armed forces. The third was the people of Afghanistan themselves, who having been fragmented by almost thirty years of violence had little unity. The fourth was regionally based narco-economies that had global implications. These dissipating forces had come together in five distinct regional units that formed solid resistance towards centralization under ATA. Figure 8.2 approximately delineates these regions.[20]

FIGURE 8.2. THE OBSTACLES TO SUCCESS IN AFGHANISTAN BY REGION

The analysis contained in "Creating a National Economy: The Path to Security and Stability in Afghanistan" provided a strategic co-ordinating mechanism and a common vision for development. Without shared vision the multitude of efforts by ATA and the international community would remain diffuse and have little impact. The paper proposed that through focusing resources in specific areas one could make the greatest impact in addressing the contemporary challenges of post-conflict Afghanistan. Furthermore, it asserted that ATA and the international community needed to combine efforts in order to transform regional illicit economies into one legitimate national economy. The best way to do this was to focus on the greatest vulnerability in each regional economy, specifically the rural areas where opium was produced.

This proposal for a focused program of development was a macro model that allowed ATA to take the lead in the re-establishment of national governance in Afghanistan. The analysis recommended using overarching planning and co-ordination mechanisms within the national policy programs, with inclusive multi-agency representation to oversee the implementation of this idea. The co-ordinating body would then be responsive to ATA and have subordinate co-ordinating committees in the districts being developed. In this manner a cohesive and unified approach, marshalling all resources, could be taken towards regional development. Furthermore, an inclusive form of command and control structure could be created by the governing authorities. It was also acknowledged that the strategy could only work if consensus in assisting ATA to choose regions for concentrated development, in selecting priority institution-building programs, and in distinguishing work requirements was created amongst all stakeholders by leadership from the principal ambassadors and supported by the heads of donor organizations. Figure 8.3 provides a visual depiction of the perceived challenges of creating common intent.

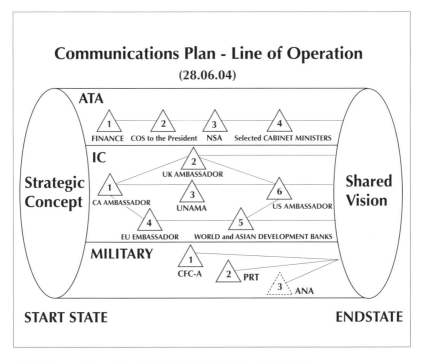

Communications Plan - Line of Operation
(28.06.04)

ATA

1 FINANCE   2 COS to the President   3 NSA   4 Selected CABINET MINISTERS

IC

2 UK AMBASSADOR
1 CA AMBASSADOR   3 UNAMA   6 US AMBASSADOR
4 EU EMBASSADOR   5 WORLD and ASIAN DEVELOPMENT BANKS

Strategic Concept

Shared Vision

MILITARY

1 CFC-A   2 PRT   3 ANA

START STATE

ENDSTATE

FIGURE 8.3. COMMUNICATIONS PLAN: LINES OF OPERATION

Figure 8.3 is a copy of the original picture that laid out initial thoughts on the creation of the network. The start state was the strategic concept, and the end state was the creation of a shared vision. The three major thrust lines, or lines of operations, were thought to be ATA, the international community and the various militaries operating in Afghanistan. The triangles represented entities or groups of individuals whose support was necessary. Although the diagram depicts linear relationships, the reality was more complex with these interconnections crossing lines of operation and having second- and third-order effects. It was also four dimensional in nature with not only geographic but temporal elements. While no common information technology infrastructure existed, people-to-people contact, low-technology communication devices (like telephones) and the Internet enabled gradual establishment of a like-minded network. More often than not, this hybrid network was a mixture of secure and non-secure or Internet systems with human interfaces. Mechanical and human sensors created shared awareness through establishing by dialogue and e-mail a common operating picture that was available to many participants, both military and non-military. Interestingly, it can be noted that while no conventional command and control mechanisms existed, the establishment of this network, united by shared intent, still provided a means of creating actions that pre-empted those of factionalized opponents.

The plan described above indicates a key difference between normal operations and current PSOs. Through high intensity or "kinetic" types of operations one attempts to overwhelm the enemy through all means available, but primarily through violence. The focus is on the destruction of the enemy rather than on consolidation of one's own forces and authority. During PSOs, such as that being conducted in Afghanistan, one must constrain the growth of the threat forces and manage the perception that there is an increase in measurable government capacity. The military component is just one element of the overall campaign. *The ultimate object is a stable and secure endstate*; in other words, "we can't leave until the state building is complete; that's why it's called an end–state."[21] As a result, during nation building, one attempts to establish and maintain a window of opportunity for as long as possible, and the objective becomes creating desired perceptions and common intent rather than destruction. It is necessary to establish, amalgamate and reinforce human and technical networks in order to create the desired effects. Unlike some of the outcomes originally proposed by adherents of network-centric warfare, it is necessary to establish hybrid networks that address the challenges of post-conflict environments, including the challenge of distributed operations that are without common technology and are affected by the exigencies of multi-national and multi-agency activities.

Furthermore, using hybrid networks can be advantageous. Van Creveld hypothesizes that the danger of information-driven command systems becomes that of distinguishing the relevant from the unimportant in the masses of information available.[22] The danger of this is borne out by recent American tactical operations in Iraq:

> Making every soldier a sensor briefs well, but when it comes to execution there are many hurdles to overcome before we get there....

> There is no staff at the battalion or brigade level capable of managing the information load we would generate at the 75%

solution. Can't say how many patrols out of the total make it to brigades and can't imagine how someone could accurately generate that statistic, but would offer that not all data is important, and tying up bandwidth burying the analysts under an avalanche of raw information is counter-productive at best. No staffs are manned to collect and process everything that comes in from the field right now, and analyzing what does come in effectively is close to impossible. That said, it is important to do the best one can to sort the wheat from the chaff. Platoon leaders and company commanders do the best processing and analysis with their own heads. Periodic assessments on their part submitted higher provide a much clearer picture than 60 patrol debriefs a week. Reports help with pattern analysis at higher HQ, but most of what comes out of that analysis we already know....

Human connectivity is every bit as important as digital connectivity.[23]

Supporting the importance of people, van Creveld focuses on the human element in command as being paramount in overcoming the inherent friction of war and winning the conflict of opposing wills. With this in mind one must view command and control in PSOs as an exercise in creating trust rather than as the more conventional idea of "imposing one's vision on the battlespace." British manoeuvre theorist Richard Simpkin articulates it simply, within the context of military forces, as a "supple chain," that is, "...a chain of trust and mutual respect running unbroken between theatre or army commander and tank or section commanders."[24] By utilizing technology to assist with articulating commander's intent, by conducting mission analysis, and by designation and use of a main effort, the decentralization of command will occur. In this setting subordinate commanders will make appropriate decisions and take action to achieve positive results without specific orders.[25] Van Creveld suggests that organizations can design their command structures to operate in the environment of chaos, or the "province of uncertainty," while less information will actually increase their likelihood of success.[26]

These theoretical ideas are borne out by practice in Afghanistan and apply to not only military organizations but all those united by common purpose. It is apparent that the human factor is dominant and that the networks established by ISAF reflect this in that they are hybrid, consisting of both technology and people. In such a setting, as described by Simpkin and van Creveld, command is predicated on dissemination of intent, creation of shared awareness, communication, and decentralized decision making.

The general trend towards the emphasis on human aspects of networks has been reflected in recent ideas proposed by the Office of Force Transformation, United States Department of Defense:

> NCW generates increased combat power by networking sensors, decision makers, and shooters to achieve shared awareness, increased speed of command, high tempo of operations, greater lethality, increased survivability, and a degree of self-synchronization.... In essence, it translates information advantage into combat power by effectively linking friendly forces within the battlespace, providing a much improved shared awareness of the situation, enabling more rapid and effective decision making at all levels of military operations, and thereby allowing for increased speed of execution. This 'network' is underpinned by information technology systems, but is exploited by the Soldiers, Sailors, Airmen, and Marines that use the network and, *at the same time, are part of it.*[27]

The emphasis on human aspects is the primary difference between the development of NCW in the Canadian and in the American circumstances. In 1998, the original concept of NCW outlined by Cebrowski and Garstka focused on emergent technology and the potential of a revolution in military affairs. For the Canadian Forces, particularly the Army, which has not had the same access as US forces to developing information management tools, the focal point has been human-centric networks incorporating technology to meet the challenges of the post–Cold War era.

During ISAF V, this effort to create networks that would promote interest and shared ownership amongst all agencies was crucial and took a great deal of work. It was recognized that without the military assisting ATA to advance this concept, both the populace and the international community would be reluctant to sustain it. The support of the international community was necessary for unified action, and without the security, planning resources and co-ordination provided by the military, ATA and the international community would be reluctant to commit resources to this strategic vision. The hybrid networks so established also provided the sensor capabilities necessary to establish a common understanding of activities transpiring in all regions of Afghanistan and bordering countries. In many ways it becomes the role of the military commander to create and sustain these networks.

In order to create agreement for the ideas contained in "Creating a National Economy," it was necessary to encourage communication, feedback and eventually accord amongst all potential participants, including ATA. Accordingly, a small group of Canadian-led ISAF officers worked diligently at facilitating common understanding and acceptance of the analysis. In light of the planned presidential and lower-house elections and likely leadership changes during the fall of 2004, it was somewhat difficult to deduce who would be in a position to become the prime proponent of this idea. However, there were a number of officials in positions within ATA at the time who could act as advocates for the concept in the foreseeable future, and these individuals were briefed on the contents of the proposal.

Simultaneously, a number of the Western ambassadors and their staffs were approached, and they demonstrated varying degrees of interest. The Canadian ambassador was informed of the initiative, and a copy of the package was sent to the Department of Foreign Affairs and International Trade for their information and further promulgation. Feedback from various international and non-governmental organizations was also solicited, and all showed support for some if not all aspects of the concept. It was also proposed that another possible venue for dissemination would be to forward the paper to consulting companies currently involved in Afghanistan for their information and to take advantage of their contacts and information. Wherever possible, the paper was distributed to consultants and contractors because they are now part of the contemporary environment of PSOs and have access to a multitude of networks.[28]

Information packages and briefings were also given to Coalition Forces Command, Afghanistan (CFC-A), to ensure that they were aware of the work that was ongoing and to gain their support.[29] The Coalition and

NATO forces in the provinces that were co-ordinating reconstruction efforts and would eventually have to act in an enlarged role were the Provincial Reconstruction Teams (PRT). They were capable of making the linkages between the officials of ATA and the international community to assist with co-ordination and capacity building. They could also act as the military representatives in a given area and speak on behalf of their commanders. In a manner similar to the joint commission observers of the NATO Stabilisation Forces in Bosnia, they would also be a "directed telescope" or a way of passing information outside the usual reporting channels.[30]

Although time did not permit before the transfer of authority from ISAF V to ISAF VI, it was thought that it would have been useful to approach the World Bank and the Asian Development Bank to solicit their opinions on a unified approach to the reconstruction of the state of Afghanistan. Numerous experiences with stability operations in the recent past have indicated the need for active involvement of donors to achieve strategic goals.

It is necessary to approach the difficult quandaries posed by PSOs in post-conflict situations holistically and identify points that must be addressed simultaneously, in a distributed fashion across elements of national power, in order to achieve the desired result. The focus of activities during the planning process, when the opponent is an ill-defined and usually a non-state actor, should not be on traditional views of overcoming threat forces, but instead should be on the attainment of the objective of the operation by linking the people and organizations necessary for success. Additionally, it is very important that all military actions be conducted in a manner that will bear public scrutiny. Indigenous participation, support and, most important, leadership of the process of reconstruction must be encouraged. Any concepts designed to assist with the renewal of wartorn societies must recognize the need for local participation in the transformation as it evolves and matures. This is particularly true in Afghanistan where the national government is the lead agency for all efforts.

Although peace should be the ultimate aim of war, unfortunately it is often not the result. In order to bring into being a truly successful post-conflict peace it becomes essential to produce societal regeneration by understanding the nature of the society being rebuilt.[31] This includes focusing on the processes that permit strengthening and on developing structures that promote networked operations. Specific military plans must be formulated that include unified and balanced efforts by all agencies to achieve shared intent and collectively promote the conditions necessary for success. Ideas such as those contained within the 3D construct will aid in this endeavour in an inter-agency milieu. Commanders and staffs must address the need to subordinate aspects of the military campaign to ensure the involvement and engagement of the international community in networked, distributed operations guided by trust, common understanding and end state. Contemporary operations in Afghanistan have highlighted the difficulties inherent in the re-establishment of a nation and confirmed the importance of these hybrid networks that not only link military and non-military activities, but also permit focused and unified activity. Only by establishing these connections can one orchestrate the elements that are required to resolve the predicament of building an enduring peace. The significance of recent Canadian experiences is just now coalescing into a recognition that they may provide the basis for a new doctrine that includes the formal utilization of network-enabled operations during interventions in post-conflict environments.[32]

NOTES

1   Observations and background regarding the creation of networked operations Afghanistan are derived from Howard G. Coombs and General Rick Hiller, "Winning the Peace in Afghanistan: Creating Effective Post-Conflict Military Intervention," *Les Lendemains des Guerres* (Kingston: Royal Military College, in press).

2   Arthur K. Cebrowski, and John J. Garstka, "Network-Centric Warfare: Its Origin and Future," *US Naval Institute Proceedings* 124, no. 1 (January 1998), 28–35.

3   The authors of this chapter are indebted to a number of Canadian Forces officers, especially Commander Chris Henderson, Lieutenant-Colonels Allen Black (now retired) and Ian Hope, as well as Majors Cathy Amponin and Mike Pepper, both of the United States Air Force, and the staff of ISAF V whose work in Afghanistan is contained throughout this section.

4   An American officer, Afghanistan, e-mail to Howard G. Coombs, July 2003.

5   Lester   Bowles   Pearson,   "Nobel   Lecture,"   11   December   1957, http://nobelprize.org/peace/laureates/1957/pearson-lecture.html.

6   For further study concerning concepts of nation rebuilding, see Michael Pugh, *Regeneration of War-Torn Societies* (New York: St. Martin's Press, 2000).

7   The most recent publication of Canadian peace-support-operations doctrine lists conflict prevention, peacemaking, peace building, traditional peacekeeping operations, complex peacekeeping operations, enforcement actions and humanitarian operations as the various categories of peace-support activ-

ities. Canada, Department of National Defence (DND), *Peace Support Operations*, B-GJ-005-307-FP-030 (6 November 2002), pp. 2-3 to 2-5. A historical perspective of the challenges experienced by Canadians during peace-support operations is contained in Desmond Morton, *A Military History of Canada: From Champlain to Kosovo*, 4th ed. (Toronto: McClelland & Stewart, 1999), 277–81.

8   It should be noted that, while in retrospect one can see UNEF I as a "traditional" or customary form of peacekeeping, for the participants this was a distinctly new type of military mission with protocols that had to be developed as the operation matured. See Captain J.A. Swettenham, "Some Impressions of UNEF, 1957 to 1958," report no. 78, Historical Section (GS) Army Headquarters, dated 2 January 1959.

9   J.L. Granatstein, "Canada and Peace-keeping Operations," report no. 4, Directorate of History, Canadian Forces Headquarters, dated 22 October 1965.

10   David Pratt, "The Way Ahead for Canadian Foreign and Defence Policy," keynote address to the 20th Annual Conference of Defence Associations Institute Seminar, available online at http://www.cda-cdai.ca/seminars/2004/pratt.htm.

11   The Drvar incident involved elements of the 1st Battalion The Royal Canadian Regiment Battle Group. A detailed discussion of these violent events, ostensibly precipitated by the resettlement of Serbians by the United Nations High Commissioner for Refugees (UNHCR) to what had become a primarily Croatian community, and their aftermath is contained in Richard M. Swain, *Neither War Nor Not War: Army Command in Europe During the Time of Peace Operations: Tasks Confronting USAREUR Commanders, 1994–2000* (Carlisle, PA: United States Army War College, Strategic Studies Institute, May 2003), 1–25. Defence analyst David Rudd has reinforced the necessity of the partnerships in the 3D concept: "The security provided by robust, well-equipped military forces in strife-torn lands opens the door to political reconstruction, which begets economic and social development, which in turn reinforces security." David Rudd, "The Canadian Institute of Strategic Studies: Commentary," available online at http://www.ciss.ca/Comment_Newpolicy.htm.

12   Pearson, "Nobel Lecture."

13   Jim Cox, "Major-General Andrew Leslie: Kabul & ISAF," *FrontLine* (September/October 2004), 8, www.frontline-canada.com.

14   "The ATA replaced the Afghan Interim Authority (AIA). In accordance with the Bonn Agreement, the ATA organized a Constitutional Loya Jirga in late 2003 to pave the way for the election of an Afghan government by early 2004." Due to a number of factors the election was delayed until October 2004 and resulted in the inauguration of President Karzai that December. "Peacebuilding in a Regional Perspective: Government of Afghanistan," http://www.cmi.no/afghanistan/background/ata.cfm.

15   Diagram from an unpublished presentation by Lieutenant-Colonel Ian Hope, "A Strategic Concept for the Development of Afghanistan," June 2004.

16   The Transitional Islamic State of Afghanistan, Ministry of Finance Consultation Draft, "National Priority Programs (NPPs): An Overview," 23/24 June 2004, 3.

17   NATO, ISAF, "Creating a National Economy: The Path to Security and Stability in Afghanistan," June 2004.

18   "Security Issues Dampen Afghan Investments," *The Associated Press*, 13 July 2004.

19   Afghanistan Ministry of Finance Consultation Draft, "Securing Afghanistan's Future: Accomplishments and the Strategic Path Forward," 29 January 2004.

20   From Hope, "A Strategic Concept for the Development of Afghanistan."

21   Lieutenant-General (now General) Rick Hillier, Commander ISAF, interview by Howard G. Coombs, 20 July 2004.

22   Van Creveld, *Command in War*, 269–70.

23   An American officer in Iraq, e-mail message to Howard G. Coombs, 28 May 2005.

24   Richard E. Simpkin, *Race to the Swift: Thoughts on Twenty-First Century Warfare*, (London: Brassey's Defence Publishers, 1985; paperback reprint, 2000), 241.

25   Ian Hope, "Manoeuvre Warfare and Directive Control: The Basis for a New Canadian

Military Doctrine, Part 2 of 2," *The Canadian Land Force Command and Staff College Quarterly Review* 5, no. 1/2 (Spring 1995), 8–9.

26   Van Creveld, *Command in War*, 269–70.

27   Emphasis added. U.S. Department of Defense, Office of Force Transformation, *The Implementation of Network-Centric Warfare* (5 January 2005), 4–5, http://www.oft.osd.mil/library/library_files/document_387_NCW_Book_LowRes.pdf.

28   "Windfalls of War: US Contractors in Afghanistan & Iraq," *The Center for Public Integrity* (Washington, DC: September 29, 2004), http://www.publicintegrity.org/wow/default.aspx.

29   Coalition Forces Command, Afghanistan, had the mission of conducting "full spectrum operations throughout the area of operations in order to establish enduring security, defeat Al Qaida/Taliban and deter the re-emergence of terrorism in Afghanistan." Untitled ISAF Liaison Officer Document (1 September 2004).

30   In the quest for certainty, commanders sometimes utilize qualified and trusted officers to act as observers and report their findings. These special agents exist outside the chain of command and report back to the originating authority in the manner of a telescope directed towards a certain point. They provide information from specified units and operations. Gary B. Griffin, *The Directed Telescope: A Traditional Element of Effective Command* (Fort Leavenworth, KS: Combat Studies Institute, 1991), 1.

31   Fuller believed that the ultimate weakness of Clausewitzian theory was its misunderstanding of the role peace played in shaping warfare, and that the violence of conflict disconnected from the strategy required for the establishment of a lasting peace resulted in nothing more than a temporary cessation of hostilities. All too often notions of creating an enduring peace are an afterthought to the achievement of a decisive victory against one's opponent. J.F.C. Fuller, *The Conduct of War, 1789–1961: A Study of the Impact of the French, Industrial, and Russian Revolutions on War and Its Conduct* (New Brunswick, NJ: Rutgers University Press, 1961; reprint, New York: Da Capo Press, 1992), 76.

32   For further reading on this topic see Allan English et al, "Beware of Putting the Cart Before the Horse: Network Enabled Operations as a Canadian Approach to Transformation," DRDC Toronto, Contract Report CR 2005-212 (19 July 2005).

CHAPTER 9

THE LOSS OF MISSION COMMAND FOR CANADIAN EXPEDITIONARY OPERATIONS: A CASUALTY OF MODERN CONFLICT?

Major-General Daniel P. Gosselin

...I will tell you that in a command-centric structure, we work to what is universally called mission command...versus risk aversion. Mission command means giving a commander a mission...giving him guidance on how you want it to proceed and, in particular, what effect you are seeking -- not detailed guidance on how to do every little thing, but what effect you are seeking...

General Rick J. Hillier
Chief of the Defence Staff[1]

A new Chief of the Defence Staff (CDS) often brings fresh perspectives to the Canadian Forces, and the recent appointment to CDS of General R.J. Hillier, a commander with extensive operational-command experience at all levels, is promising a different approach to command in the CF. Indeed, within a few weeks of taking command of the Forces in February 2005, General Hillier implied several times that it was time to adopt a "mission command" philosophy of leadership. At an April 2005 meeting of the Armed Forces Council, the senior military body of the CF, the CDS made the point that "a 'mission command,' vice risk-averse approach, is required for the oversight of [CF] operations,"[2] strongly implying that he intends to let commanders perform their responsibilities with this time-honoured doctrine as the core concept for commanding and decision making.

The philosophy of mission command, with fundamentals such as unity of effort, freedom of action, trust, mutual understanding and timely decision making at its core, focuses on "telling subordinates what to do, not how to do it."[3] But the challenges that the CDS will face in his attempt to revamp the leadership practices currently espoused for Canadian expeditionary operations will be many and will require both an overhaul of the CF command system and the development of new attitudes about decision making on the part of Canadian strategic military leaders and politicians.

In the past ten years mission-oriented command — the concept that subordinate commanders are given wide latitude and use their initiative and creativity to achieve strategic and operational goals — has for all intents and purposes disappeared from the CF, particularly for Canadian operational commanders involved in international operations. Deployed commanders nowadays are delegated limited authority to fulfil their responsibilities,[4] and their role is largely restricted to one of senior Canadian administrator in theatre addressing national command issues, with most key decisions elevated to the strategic headquarters in Ottawa.[5] This experience is certainly not only unique to Canada; indeed, many military analysts, scholars and practitioners of war have recognized that the nature of decision making for military operations has changed significantly in recent years,[6] and a number of Canada's allies have noted the same trend.[7]

The rapid rise and pervasiveness of information technologies and the exponential growth of high-speed global communications are often identified as the elements that have most contributed to altering the nature of military decision making, affecting the speed at which decisions are taken, and by whom and where those decisions are taken.[8] The changes in the nature of military decision making have even led some authors to question the continued relevance of the classical doctrine of three levels of war,[9] and for others to propose a thorough re-examination of the industrial-age hierarchy that defines the military.[10] Many military analysts are arguing that in the new type of modern war being fought, the distinctions between the strategic, operational and tactical levels of war are becoming blurred, contributing significantly to the complexity of the war and to the emergence of new patterns of decision making.[11] The Canadian military is also facing its own crisis of excessive centralized controls, often dominated by the staff matrix, leading to the recent statements by the new CDS to adopt mission command.

This chapter examines decision making for Canadian expeditionary operations and its impact on mission command in the CF. It contends that a continuing and irreversible trend in modern conflict during the past decade has been the increased compression of the levels of war and that a consequence of this compression is increased centralization of CF decision making and a corresponding decrease in the use of mission command. The nature of expeditionary operations in which the CF expects to be engaged in the future, combined with the risk-averse decision-making culture that exists at all levels of decision making within the CF and in government, means that Canadian operational commanders can expect to continue to be subjected to restrictive controls in the performance of their responsibilities. For the CF, the implications of this condition warrant a serious questioning of the continued relevance of the need for a Canadian operational commander as envisaged in Canadian doctrine.

Mission Command and Its Evolution

> *Mission command:* The CF philosophy of command, which basically relies on a clear understanding of the commander's intent to co-ordinate the actions of subordinate commanders and which thereby allows them maximum of freedom of action in how they accomplish their missions.
>
> *Leadership in the Canadian Forces: Conceptual Foundations*[12]

"Mission command…is robust and must endure," stated a recent UK military concept publication outlining a long-term vision of the way in which British forces and their methods are expected to develop at the strategic and operational levels in the years ahead.[13] In the same vein, the newly released series of manuals with the title *Leadership in the Canadian Forces*, issued under the authority of the CDS, reaffirm that mission command is the CF philosophy of command, which "explicitly recognizes the necessity of allowing subordinates maximum freedom of action consistent with commander intent."[14] Understanding mission command in the context of C^2 in the information age is central to appreciating the impact of this philosophy on strategic and operational decision making in Canada.

The evolution and theories of command methods and C^2 — from concepts to systems — is a broad topic with many sub-themes, and a comprehensive review is clearly beyond the scope of this chapter. The area of interest most relevant to the argument presented here centres on the nature of command and control at the vital operational-level and strategic-level interface, the key node of interaction between national policy and direction and tactical execution. Indeed, success in modern military operations most often centres on effective and robust command.

Thomas J. Czerwinski, a systems analysis expert who served in both the Marine Corps and the US Army, outlined three methods of command that commanders have employed over the past centuries to achieve their missions: command by direction, command by plan and command by influence. Command by direction is the oldest method of command and virtually the sole method until the middle of the eighteenth century. Until then, commanders could still see the entire battlefield and could either attach themselves to one element of the force judged to be decisive in the battle or move from unit to unit as the situation dictated. They could essentially direct their forces all of the time.[15] Technological advances have been used over time to allow commanders to always "see" the entire battlefield and command their troops. While in the 1700s Frederick the Great used the telescope to establish his headquarters at a fixed location overlooking the battlefield rather than having to rush around it,[16] General Tommy Franks, Commander Central Command, attempted to emulate this approach from his command centre in Tampa, Florida, using today's technology to conduct the war in Afghanistan.[17]

The advent of firearms, the revolution of tactics and organizations, and the increase in size of armies "meant that battlefield fronts frequently extended to the point they no longer formed single, coherent wholes that could be controlled by one man." In recognition of these changes and the growing difficulties of directing forces over a larger battlefield, the command-by-plan method appeared, and consists of a highly centralized method of "trying to plan every move in advance, relying on highly trained troops and strict discipline to carry out the scheme as ordered."[18] The contemporary doctrine of concentrating on identifying and neutralizing centres of gravity in a campaign, for example, is a key characteristic of this method, which trades flexibility for operational focus.[19] Many elements of this method of command remain in place today, with the air tasking order being perhaps the most evident.

By 1870, tactical flexibility was needed to deal with the confusion and uncertainty of the large battlefields — owing to the orchestration of large forces in time and space — and a "new" method of command, which relied on "a decentralization of command proceeding from the top down," appeared and involved "in particular a delegation of responsibility to company commanders."[20] Command by influence, or mission command as it is most often referred to these days, offered new degrees of flexibility in carrying out tactical actions, a necessity to account for the dispersion and mobility characteristics of the "modern" battlefield.

Field Marshal Helmut von Moltke, Chief of the General Staff of the Prussian Army from 1857 to 1888, played a decisive role in the development of *Auftragstaktik*, mission-oriented command,[21] which focuses on fostering independent thinking and acting among subordinates. He explained that "diverse are the situations under which an officer has to act on the basis of his own view of the situation. It would be wrong if he had to wait for orders at times when no orders can be given. But most productive are his actions when he acts within the frame-work of his senior commander's intent."[22] As Moltke noted, *Auftragstaktik* is possible because of one key principle at the heart of the concept: "the subordinate is to act within the guidelines of his superior's intent. Knowing his superior's intent, the subordinate thus works toward achieving it."[23] Unlike other forms of command, this method takes "disorder in stride" and considers the chaotic war environment "as inevitable and even, insofar as it affected the enemy as well, desirable."[24] Thus, mission command places greater reliance on the initiative of subordinates based on local situation awareness, "which translate to lowered decision thresholds. It relies on self-contained…units capable of semi-autonomous actions," occurring within the limitations determined by the concept of operation established and consistent with the higher commander's intent.[25]

In his seminal work *Command in War*, van Creveld reviewed the historical evolution of command systems and the way in which such systems operated. He remarked that history had abundantly demonstrat-ed that the most successful armies have been those who "did not turn their troops into automatons, did not attempt to control everything from the top, and allowed subordinate commanders considerable latitude."[26] He further acknowledged that uncertainty is the central fact that all command systems have to deal with, and it has always been a decisive element in determining the structure of command.[27] It is this uncertainty, unpredictability and confusion — inevitable in war — which tests command systems and "forces" them to evolve over the years. In many ways, command and control is a continuous process of uncertainty reduction.[28]

In van Creveld's opinion, however, there are two basic ways of coping with this uncertainty: centralization *and* decentralization. He contends that centralization and decentralization are not "so much opposed to each other as perversely interlocking,"[29] and adds that in war "to raise decision thresholds and reduce the initiative and self-containment of subordinate units is to limit the latter's ability to cope on their own and thus increase the immediate risk with which they are faced." Over-centralization, which often equates with over-control, can lead to indecisiveness and loss of operational tempo, and, therefore a greater likelihood that strategic decisions will be taken without an appreciation of the context of the situation on the ground, leading to distrust of superior commanders. Moreover, removing from tactical and operational leaders a "sense of responsibility, ownership, and empowerment decreases motivation, retards creative thinking and problem solving, and results generally in less effective execution."[30]

It is clear that looking at systems of command in isolation — that is, command by direction, by plan or by influence, or any combinations thereof — is too one-dimensional as a command framework to deal with today's operations, especially at the strategic and operational levels of war. Mission command, as originally expressed by Moltke, appeared more suited when there were only two levels of war, the strategic and the tactical. While network-centric warfare (NCW)[31] may eventually deliver on its promises and make possible the achievement of the principles of Moltke's mission command, C^2 is more complex nowadays and therefore requires a more innovative conceptual approach with which to analyze it. The recent work of two Canadian researchers who took on the task of re-defining command and control provides some interesting new directions in this regard.

Ross Pigeau and Carol McCann, researchers with DND's Defence Research and Development Canada, took a new angle to analyze C^2 and "decided to start from scratch and re-conceptualize the whole area" of command, control and C^2, with the fundamental underlying assumption that "only humans command."[32] Their purpose was to capture the essence behind the terms *command* and *control* and not take the individual words or the expression C^2 as granted. Their definition of command stresses the "fundamental assumption to which operational commanders have alluded to time and again — that humans bear the burden of command," because commanders, not systems, demonstrate the range of innovative and flexible thinking necessary to solve complicated and unexpected problems. The authors define three dimensions of command — competency, authority and responsibility — with the achievement of a maximal, or balanced, envelope of those dimensions being central to effective command. If one of those three elements is weak or missing, an imbalance exists and the possibility of command failure or poor performance is significantly increased.

Many military analysts argue that the crux of the issue in devising contemporary command systems is in determining the appropriate balance between centralized control and decentralized execution. The Canadian researchers do not accept this one-dimensional explanation; for Pigeau and McCann, *command* and *control* are complementary, with control always subordinate to command. The authors define *command* as "the creative expression of human will necessary to accomplish the mission," while *control* is defined as "those structures and processes devised by command to enable it and manage risk."[33] In this context, mission command is thus a philosophy intended to maximize human creativity, initiative and diligence.[34] While van Creveld's analysis centred on the growth and evolution of command systems to deal with the fog of war, the Canadian authors add clarity to this line of reasoning, explaining that the management of risk is a key factor influencing the development of control structures and processes. While uncertainty and battlefield chaos certainly contribute significantly to increasing operational risk and potential mission failure, Pigeau and McCann point out that many other elements contribute to it. In the end, however, it is command, and only command, that creates and changes the structures and processes of control to suit the uncertain military situations.[35]

Until recently, command systems had evolved to allow for the greatest possible decentralization — initiative, flexibility, decision making, situational responsiveness, and execution at the lowest possible level — with sufficient centralization to enable superior commanders (and headquarters) to retain control over subordinate ones. While it is true that new NCW concepts, developed to encourage self-synchronization and co-ordinated effort across all levels, are designed to promote mission command, it is also true that the networks created to enable mission command facilitate the centralization of decision making and encourage over-controlling behaviour. One American author, highly critical of the involvement of senior commanders in tactical matters, goes as far as to argue that "...the most serious problem in the US military today is the continued deterioration of the previously successful and well proven method of centralized direction and decentralized execution of planning military actions at all levels," and recommends that this trend be arrested by adopting mission command "top to bottom."[36]

To many commanders and military analysts, adopting a mission-command philosophy is definitely the only sensible approach to address the challenges that C^2 faces today.[37] In fairness, espousing Moltke's mission-command philosophy is likely too simplistic a solution to solve all the command problems highlighted above and fails to account for the complexity of military operations, the dynamic politico-military environment that sometimes exists, and the elaborate command structures that have been constructed to deal with this complexity. Pigeau and McCann perceptively remark that the important point is to find the correct balance between encouraging "creative command and controlling command creativity."[38] This is precisely the dilemma that faces the CF today, especially at the strategic-operational interface, and the challenge is therefore for commanders at all levels to find the right dosage of mission-command principles to guide them in establishing the C^2 framework for expeditionary operations.

Canadian Strategic- and Operational-Level Command

General R.R. Henault, then the Deputy Chief of the Defence Staff (DCDS) and the senior CF officer responsible for all CF expeditionary operations, stated a few years ago, upon reflecting on the lessons learned from the Canadian participation in the 1999 Kosovo air campaign, that "it is clear in many ways that the role of the strategic commander has changed *considerably* since the end of the Cold War."[39] He could have pursued his analysis by adding that it is equally clear that the role of the Canadian operational and tactical commanders has also changed measurably in that period. Therefore, this section of the chapter reviews Canadian strategic and operational command during recent expeditionary operations.

The CF joint doctrine that provides the fundamental principles by which all CF operations are conducted has progressed significantly in the past fifteen years. Many CF operations — domestic or international, as part of a coalition or an alliance — have provided a rich collection of lessons that have shaped the CF operational doctrine.[40] The doctrine for C^2 of operations is now well tested and solidly entrenched. Task forces, or joint task forces when two or more environments participate in the same operation, are constituted as soon as a mission is launched, with the designated task force commander reporting to the CDS. For contingency international operations, command of combat units and support elements is vested in a joint task force (JTF) commander appointed by the CDS.[41] As soon as the CDS establishes a JTF, a separate chain of command is activated with the appointed JTF commander reporting to the CDS. In this case, the CDS is the Canadian strategic commander and the JTF commander is designated the operational commander.[42]

Upon appointment, task force commanders are provided with terms of reference and a statement of the CDS's intent. The commander of each Canadian rotation of a mission will receive CDS-issued documents, which will usually contain explicit Canadian government intent for the campaign and the operation; national strategic objectives; constraints; military objectives; the CDS's intent; a description of the resources available to execute the mission; an elaboration of the command and control arrangements; and the delegation of authority for the administration, support and discipline of all deployed CF elements and personnel.

To ensure there is no potential misunderstanding in interpreting the terms of reference assigned, all designated Canadian operational commanders are interviewed by both the CDS and the DCDS before departure from Canada, and for the larger or more contentious missions, designated commanders will frequently meet with key personnel from the Privy Council Office and Foreign Affairs Canada as well. Thus, as General Henault claims, "the mandate of the operational commander [is] clear and set[s] the foundation of the command and control relationship."[43] Indeed, most, if not all, elements required to facilitate the expression of mission command between the Canadian strategic and operational levels appear in place: competent and qualified operational commanders, clear guidance and direction received from the senior strategic commanders, and a common understanding of the Canadian objectives and the CDS intent for the mission.

In practice, however, limited mission command takes place in the Canadian expeditionary command structure. Canadian operational commanders are given limited opportunity to exercise their assigned responsibilities since their headquarters (commonly referred to as the Canadian *national command element* [NCE]) are seldom equipped or staffed with the robustness necessary to perform the functions expected by NDHQ. In today's context for military operations, Canadian task-force elements are usually operationally subordinated to the component commander of an alliance or coalition, with "hard" operations matters left to Canadian tactical commanders to sort out with their respective component operational commander. Thus, the Canadian operational task force commander is usually not functioning as an operational-level commander and not doing bona fide operational activities, such as integrating operational intelligence and planning and conducting operational movement, manoeuvre, firepower, support, and force protection; he is not fully exercising operational C^2 over assigned forces.[44] In addition, because the NCE is often cobbled together from disparate organizations, it usually possesses limited capability to perform the necessary oversight of operational matters taking place within the task force, with the result that most operational decisions are elevated to the strategic commander in Ottawa.

With most key decision making taking place in Ottawa, strategic commanders in recent years have thus spared no effort to improve situational awareness at NDHQ to facilitate this decision making, to be in a position to better identify and assess the strategic and operational risks, and to make sound recommendations to the politicians. As General Henault admitted a few years ago, "it is imperative that the strategic commander have the means and tools available to remain current on the intelligence and operational situation on the ground."[45] In fact, many improvements in this regard have taken place in recent years, and more are planned for the future.

The strategic headquarters, NDHQ, is now endowed with a small but highly professional military joint staff and a state-of-the art command centre to assist the CDS in performing his duties. In the post-9/11 era, the senior CF leaders have been pursuing improvements to the intelligence networks and the surveillance and reconnaissance capabilities, and to the CF Command System to expand the capabilities of the secure joint C^2 information system in order to make it more joint and to bring to the strategic headquarters a "common operating picture" for commanders. General Henault admits that "the upgrades to NDCC [National Defence Command Centre] were...a step in the right direction to help accommodate the technology that is required for the strategic commander to *maintain command and control*."[46] While significant improvements have taken place at the strategic level for the planning, directing and controlling of operations, the Canadian operational-level doctrine and the command construct have not kept pace with the progress; as a result, there remains a significant shortfall at this level, which affects the Canadian strategic/operational-level interactions. For one, the Canadian strategic commander now possesses more "tools" and a better "common operating picture" than does the deployed operational commander, clearly facilitating and encouraging "intrusions" at the tactical level.

But it is not solely the fragility of the NCE that contributes to more centralization of Canadian decision making during military operations; indeed, if this were the case, it would be a simple matter of adding the necessary robustness, expertise and depth to the Canadian operational-level command headquarters to counterbalance the continued urge to centralize. The reality is that Canadian strategic commanders are reluctant to delegate authority to task force commanders. Changes to the scope of the Canadian mission — even of a small nature — to the rules of engagement, approvals of targets, and most administrative, contracting and financial authorities are centrally retained, with the result that Canadian national commanders are usually limited to performing a military liaison role and to co-ordinating relatively minor administrative tasks for the Canadian task force. Consequently, in light of the changing relationship between the strategic and operational commanders, is it even realistic to advocate adopting mission command in the CF for expeditionary operations, when the environment persistently encourages centralization of decision making? Or is the role of the Canadian operational commander, as envisaged in the current joint doctrine, about to fade away? The next section will provide some answers to these questions.

Modern Conflict and Centralized Decision Making

Douglas Bland, a Canadian military analyst researching civil-military relations, remarked a few years ago that "the history of warfare and civil-military relations in liberal democracies...has been a steady advance in the construction of machinery to allow continual political direction and control of military activities and decisions in the interest of the state."[47] This section discusses six elements that continue to contribute to increased centralization of decision making in Canada and offers reasons why many Canadian operational issues will continue to rise to the strategic level. These elements are the continued compression of the levels of war; the complexity of military operations in which Canada is expected to engage in the future; the character of civil-military relations in Canada; the influence of the media in a democracy; the legacy of the Somalia mission; and the pressure imposed by international law.

Compressing the Levels of War. In his remarks to the Standing Committee on National Defence and Veterans Affairs on mission command in May 2005, General Hillier implied that this new leadership philosophy would mean an acceptance of greater risk in decision making, leading to a reduction of control mechanisms imposed on operational and tactical commanders. Commanders at all levels would henceforth be delegated more authority and given greater flexibility of action. In essence, risk would be managed at the lower levels in the chain of command.

Discussions in recent years about decision making in modern conflict have frequently focused on a new time and space relationship between the classical levels of war (strategic, operational and tactical), where, until recently, activities have tended to be conveniently aligned along these three tiers for planning and decision-making purposes. The continued compression of the levels of war in recent years is certainly one of the most serious obstacles to the achievement of a mission-command philosophy.

Military analysts and doctrine writers all acknowledge, explicitly or implicitly, that some important transformation is happening to the levels of war and their interaction, creating a "compression" and "blurring" of the levels, which significantly affects decision making. The Canadian key-stone joint-doctrine manual, *CF Operations*, considers this compression and blurring to be a reality of modern military operations and, as a solution, stresses the importance of co-ordinating activities at all levels to ensure successful mission accomplishment. Implied in Canadian doctrine, therefore, is a recognition that decisions that used to be taken at one level of war are, by necessity or by choice, now taken at a different level.[48] The lines separating the three levels are more difficult to discern because military activities end up being performed at different levels.

To General Wesley Clark, NATO Commander and Supreme Allied Commander Europe during the 1999 Kosovo War, this compression of the levels of war is driven by the unique nature of strategic decision making demanded by modern conflict. With its high political stakes, Clark is adamant that given the types of complex operations being conducted these days, strategic commanders *must* be involved in the decision making that would normally be left to operational commanders. The fact that the levels of war are compressing means a more active and direct role by strategic commanders, Clark adds. Consequently, a senior commander like himself needs to "have a strong grasp of detail" to be able to take those day-to-day decisions that he or she must now assume.[49] The outcome of this centralist command approach meant that Clark was constantly seeking information from lower-level operational and tactical commanders to satisfy his information needs.[50]

Others have argued a different cause and effect to this compression of the levels of war. The preponderance of military analysts contend it is the advent of global high-speed communications that has now given higher commanders the tools, unavailable before, to take decisions that until then were taken by lower-level commanders. Articles abound in recent years supporting this theme, "blaming" strategic commanders for getting involved in what many consider to be low-level decisions and for introducing excessive controls during operations — simply because the means are now available. One military author sums up best this line of reasoning: "Advances in communications allow senior leaders to observe events in near real time from thousand of miles away. This promotes a false impression that remote headquarters can perceive the situation better than tactical commanders on the scene. Consequently, not only must tactical commanders report to operational commanders, but the latter often issue orders to tactical commanders."[51] The result is that intermediate commanders — theatre and operational — are bypassed and relegated to being information administrators, as strategic commanders immerse themselves in details.

It can be argued that the growth of information networks is the main cause of this trend. Evidence suggests that networking, which supposedly promises decentralization and affords greater initiative to subordinates, creates the opposite effect because "theatre commanders increasingly use information technology to make decisions that would normally be the province of tactical commanders."[52] A recent RAND Corporation study on network-centric warfare confirms that recent operations in Afghanistan and Iraq indicate that "networking is reinforcing the tendency toward centralized C^2 within the US military and it thus not delivering the benefits of decentralized decisionmaking."[53] Technology of the type now available to commanders can be a "two-edged sword, especially when developments lend themselves to ever greater centralization and, in extreme cases, to battlefield micromanagement."[54] During Operation Enduring Freedom, the U.S. campaign against terrorism in southwest Asia in the aftermath of 9/11, senior leaders in Central Command "not only observed but also second-guessed subordinate commanders" and exercised direct command in real time over forces in Afghanistan from the Central Command headquarters in Florida.[55] Lieutenant-General Michael DeLong, second-in-command of Central Command for the period 2001–2003, proudly provides a vivid description of this capability:

> [W]e were set up to monitor this war with some of the most sophisticated technology in the world: high-tech, flat-screen plasma TV sets covering the action from GPS devices in most of our major military equipment and multi-million-dollar unmanned drones. It would be a spectacular combination of the old and the new, of ancient warfare and modern technology. It would be truly a new type of war.[56]

There is no doubt that the emergence of the NCW concept is profoundly influencing C^2 at all levels and also accelerating the compression of the decision-making cycles. But the opinions of military analysts and practitioners on what NCW will mean for future C^2 and decision making are quite diverse. Admiral Owens, a former vice-chairman of the U.S. Joint Chiefs of Staff, represented the views of many a few years ago when he argued that the new communications systems are of "profound significance...with regard to traditional concepts of military command, control, hierarchy, and organizational structure," and they will allow and prompt a much broader diffusion of knowledge that is relevant to the demands to combat "throughout all levels."[57] He implied that this C^2 revolution will offer possibilities for greater centralization of decision making and a compression of the chain of command: "...in many places the chain of command would be replaced by secure and powerful networks that relay commands and critical battlespace information from the area of conflict to key decision makers, and from leaders...to the combatants."[58]

Canadian military analyst Elinor Sloan also believes that "advanced military technology will...allow soldiers to know as much about the battlefield as the generals."[59] Self-synchronization will increase the opportunity for tactical forces to operate "nearly autonomously and to re-task themselves through exploitation of shared awareness and the commander's intent."[60] For Sloan, however, this does not necessarily mean increased centralization of decision making; rather, this will mean a "delay-ering" of military organizations and the introduction of new command pro-tocols to give more initiative and empowerment to the local commander at the expense of the theatre-level commander.[61] The advent of NCW may compress operations and levels of war, but shared awareness and self-syn-chronization will offer new opportunities to adopt mission command.

The emergence of the "strategic corporal" is another factor that tends to elevate issues and decisions to higher levels.[62] Not only are politicians and senior commanders becoming more tactical, but also the soldier is becoming a strategic player in modern conflict, especially when participating in convoluted peace-support operations where, in contrast to war, his or her actions can have strategic significance.[63] International media can broadcast the actions of soldiers in combat and stability operations and their demeanour in humanitarian operations in minutes and affect domestic politics and even the course of an operation.[64] Moreover, military operations have diffused across all levels of war so that actions at one level will have a direct effect at another. Consequently, it is reasonable to expect that political leadership will maintain a high interest in the developing situations in a theatre of operations to ensure its strategic objectives are met and national interests respected.

The central theme that emerges from this brief review of the compression of the levels of war is that a significant and influential change is taking place in the nature of decision making in modern conflict. Decisions that used to be considered within the purview of commanders at one level of war are not any more, with the consequence that the role of commanders at every level is changing dramatically. In practice, however, this compression has meant in recent years an increased centralization of planning and direction and, by extension, more involvement by strategic commanders in operational and tactical matters. There is a reluctance to delegate authority, and controls which did not exist until recently are now part of the day-to-day environment of operational and tactical commanders. This progressive trend of more centralization, limited authority and over-control is certainly present in Canada.

Complexity of Operations and Military Autonomy. Decentralization/ centralization of decision making for military operations is conditioned to a large extent by the type and complexity of the operations the military is engaged in, which by extension is a key determinant of the degree of involvement by both Canadian political leaders and strategic commanders. The trend of the past fifteen years confirms that Canada will not become involved in conventional high-intensity war fighting and instead will concentrate on a range of operations that will cover the low- to mid-intensity level.[65] The recent CF vision outlined by the government in the April 2005 Defence Policy Statement speaks of greater CF involvement in operations in "failed and failing states," with these missions now being "far more complex and dangerous" than before and taking place in a complex and chaotic environment.[66] Based on the types of missions the CF will be engaged in, centralization of decision making is not expected to go away.

One military analyst, Martin L. Cook, correctly points out that only in large-scale warfare on the order of the 1991 Persian Gulf War or the

Operation Iraqi Freedom campaign of the recent Iraq War will political leaders be likely to give the military a large measure of autonomy in conducting military operations. Cook takes this argument further, arguing that "the lower one goes on the scale of contingencies in peace and humanitarian operations, the greater the complexity one can expect in the intermingling of political ends sought, concerns for domestic political support, issues of media coverage and public reaction to it (the so-called 'CNN effect'), and the military means employed."[67] While Cook's linear extrapolation that humanitarian operations are more complex than full-scale war is clearly too simplistic to accept — and certainly does not represent today's realities — it remains that peace-support operations of the Somalia, Kosovo and Afghanistan types (the International Security Assistance Force phase, not Operation Enduring Freedom) are certainly the most complex types of operations in which the military can be engaged, with important implications for decision making.

Peace-support operations are usually carried out in large coalitions (UN- or US-led), which add to the complexity in two important ways. First, the formulation of coalition strategy, the harmonization of operational issues and the co-ordination of national support require much effort; these discussions and negotiations are typically convoluted and frustrating and, as a result, are often carried out at the national military and diplomatic levels.[68] General Henault highlights some of the challenges of coalition warfare: "multinational coalitions will result in all participating nations having to continue to face both the ethical challenges of dealing with other nations and the political challenges of finding consensus among many partners and agendas."[69] Second, the requirement to maintain strong and "indivisible" Canadian national command in an era when coalition C^2 arrangements are often complicated tends to quickly elevate many discussions about the scope of the mission, tasks assigned to Canada, command reporting relationships and even key logistical arrangements (such as food services and medical) to the strategic level.[70] As a result, this situation usually leaves the Canadian deployed commander limited to a senior liaison role, with most negotiations driven by NDHQ staff, and all key decisions taken in Ottawa.

Many factors render modern peace-support operations fairly complex, all of which contribute to removing decision making from the Canadian operational commander. As one recent report states, the "complexity of the new peace-support efforts is due not only to the increase of activities" that occur for these types of operations "but also the diversity of the groups that participate in them." The authors of the study identify over thirty variables that can contribute to the complexity of a peace-support operation, with the most significant variables being number of participants and national actors, number and diversity of agencies, complexity and stability of organizational structure, inconsistency of organizational procedures, number of communication channels influencing a military commander, level of adversary violence and asymmetry, complexity of the environment and terrain, level of public fear, level of restraint required and level of time pressure.[71] Thus, peace-support operations of the three-block war scenario type are inherently more complex, more ambiguous and riskier than humanitarian operations.

Figure 9.1 offers a more elaborate, yet still simplified, presentation of the variation of the degree of military autonomy in a range of modern military operations, from the humanitarian assistance version to full-scale combat operations. It can be seen that the measure of military autonomy is at its minimum in the most complex conflicts, such as a three-block war scenario. Those conflicts often have high political stakes (Kosovo), may involve the use of deadly force around civilian populations (special forces operations in urban Afghanistan), include many civilian agencies, international non-governmental agencies and media outlets operating on the ground (Yugoslavia), and are of high interest at home (Haiti). This is precisely the type of operations in which Canada has expressed the strongest interest to engage in the future.[72]

The consequences of this interest for operational commanders are twofold. For one, this type of operation requires the most intense co-ordination in Ottawa to achieve an integrated strategy that draws on Canada's diplomatic, development and defence resources, thus moving many elements of military co-ordination and decision making away from the operational level and to NDHQ and other departments and agencies in Ottawa.[73] The centralization of decision making is exacerbated by the fact that the Canadian government has specifically committed to "pursue an integrated strategy that draws on Canada's diplomatic, development and defence resources," which includes "a central role for the Canadian

Forces."[74] Second, because of the greater visibility given to tactical actions on the ground (the "strategic corporal" effect) — as compared to conventional war fighting — additional controls are imposed on the tactical and operational commanders, and their authority is more limited. As one leading authority on decision making points out, the typical reaction to complexity and uncertainty is to increase controls and to increase centralization of decision making.[75] In short, all these elements contribute to minimizing the role of the Canadian operational commander at the expense of the strategic commander.

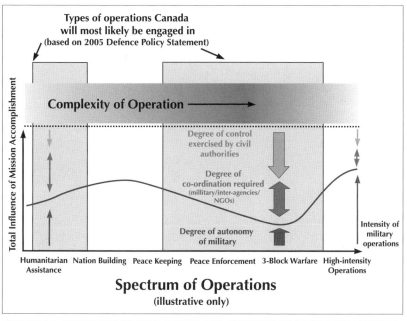

FIGURE 9.1. SPECTRUM OF OPERATIONS AND COMPLEXITY (FIGURE BY AUTHOR)

Civil Control of Military Operations. The involvement of politicians and civilian considerations in the military sphere during the conduct of operations is not about to melt away, and their involvement will invariably mean more active decision making on their part.[76] History is full of examples of politicians engaging in the management of the conduct of a war, like President Abraham Lincoln during the Civil War and President Franklin Roosevelt and Prime Minister Winston Churchill during the Second World War. President Lyndon B. Johnson and Secretary of Defense Robert McNamara became notorious for their micro-management of the bombing campaigns against North Vietnam.[77] Visions of political leaders engaging in what some may consider as military-only decisions, such as choosing target sets and prescribing rules of engagement, continue to haunt military leaders. As one author recently noted, "Often, this is captured in the wish that political leaders would wholly restrict themselves to making the decision to use force in the first place and then get out of the way and 'let the professionals' (i.e., military officers) determine the aspects of the operational plan."[78] This wish certainly contains a degree of wisdom insofar as it cautions against unqualified individuals making military operational decisions; it is certainly consistent with Samuel P. Huntington's classical theory of civil-military relations, which advocated a more rigid separation between the military and the politicians.[79] However, this view of civil-military relations represents an idealized concept that does not correspond to today's realities.[80]

The increased exposure of the armed forces to various media, the perceived greater accountability of politicians for a country's military actions, the blurring of the "front" and "rear," and of "war" and "peace," the changing views on the use of force, the participation of military specialists in controversial missions, and the sensitivities and politics associated with targeting and potential collateral damage[81] are all adding to create fundamental and far-reaching changes to civil control of the military and the national command requirement for military operations.[82] General Henault, writing in 2000 in his capacity as DCDS and reflecting upon the 1999 Kosovo air campaign, admitted that "...the unique characteristics of the type of conflicts that will dominate the future, when combined with the demands of the international media, will continue to result in unprecedented scrutiny of any military action."[83] In the post-9/11 era, this scrutiny has been accentuated in Canada.

It is fair to state that politicians — and their staffs — over the years have generally not been confident that the military field commanders have had the requisite sensitivity to the political aims and implications of any military action or activity. The words of Prime Minister Mackenzie King are instructive in this regard and a useful reminder of how politicians periodically stereotype military advice. King complained that Minister of National Defence J.L. Ralston often stood up for the generals and fought the cabinet on their behalf. King eventually said of Ralston: "I have talked to him again and again. I have asked him not once but many times why he does not tell the generals what we, in cabinet, think instead of continually telling us what the generals think. Generals are invariably wrong."[84] Desmond Morton, a Canadian historian, maintains that his conversations with government backbenchers sitting on parliamentary committees revealed a mistrust of the military commanders who had come before them.[85]

In truth, politicians have been proven right on several occasions in the Canadian context when military commanders were "naïve" about the political consequences of military actions or rationales.[86] In today's environment there is always concern that the field commander will not exert the degree of control over his forces to "ensure that attainment of the political objectives is never compromised" — as happened with the Canadian Airborne Regiment during the 1992 Somalia operation.[87] The situation with General Clark over the use of ground forces and strategic targets in Kosovo, resulting in his dismissal from his NATO post, is indicative of the seriousness that politicians attach to those matters. With politicians concerned that military commanders may not appreciate the political nuances of a developing situation, micromanagement mentalities develop at the strategic level, direct interference with the field commanders is more frequent, and restrictive controls are put in place. As some military analysts recently noted, "[b]ecause of the enhanced political content of conflict in a world of instantaneous information, centralized execution will often accompany centralized control,"[88] restricting significantly the autonomy of the operational commander.

In Canada this accrued political interest is also reflected in the growing role of the government's central agencies in military affairs, such as the Privy Council Office and the Prime Minister's Office, which contributes to greater engagement by the minister of national defence and the prime minister. While many would argue that the Canadian government may have been largely indifferent about defence in the 1990s,[89] at least until the Somalia incident came to light, the post-9/11 period is seeing a much more active role in defence matters by elected officials, especially for overseas operations. Ironically, some officers who had been advocating for years for greater political interest in defence matters are now complaining of "intrusion" and micromanagement. In the end, "war is ultimately about politics, and civilian control of the military is in the democratic tradition,"[90] and this responsibility will continue to be exercised by the politicians as they see fit.

Continual political direction and control can certainly be expected to be
more pronounced in times of crisis or conflict, placing even more
pressure on the military command system and strategic military
commanders, but if the last few years are any indication, Canadian
politicians will remain actively engaged in the defence debate even if a
crisis is not looming. This will mean that strategic military commanders
will likely continue to exercise tight control of CF overseas operations to
meet the needs of the political masters. Until politicians become
comfortable that military commanders' decisions will be taken to meet the
civil authority's political objectives without undue risk to the missions
and subordinates, the situation is unlikely to change in the near term. [91]

The Influence of the Media. The advent of real-time news coverage in the
past decades has led to unmatched awareness of military operations as
they unfold and to a continuous scrutiny of strategic decisions. The
flurry of political and military activity that can result from what may be
perceived at the operational and tactical levels as "minor" can be
staggering. A few incidents in the past years involving the CF have confirmed that the "CNN effect" has also reached Ottawa, not just
Washington.[92] For instance, during one of the operations of the battle
group based on the 3rd Battalion, Princess Patricia's Canadian Light
Infantry, in the region of the Tora Bora Caves in Afghanistan in 2002, one
platoon was tasked to provide security to US criminal investigators who
had stumbled across recently dug graves; the speculation was that Osama
bin Laden might have been buried at this site as a result of the bombing
in the region. An embedded reporter equipped with a video camera
accompanied the platoon. While NDHQ staff were told that Canadian
soldiers would only provide security and would have no role in the
exhuming of the bodies, the investigators eventually requested assistance
from the Canadians. For a number of reasons NDHQ was originally
unaware of this change of plan, and as events would unfold, the reporter
filed his footage, immediately flashing it on television screens across the
country: it showed Canadian soldiers helping US civilian investigators
pulling bodies out of the graves. The flurry of political and military activity that resulted from this task was clearly out of proportion to the risks
taken by the soldiers — a task that, incidentally, had been approved by the
senior coalition operational commander at Central Command. Without a
doubt, we are living in the information age where the media, the Internet
and the flash-news mentality are making the world a very small place.[93]

Technological advancements in the broadcast arena, combined with the public's increased thirst for news, result in an even higher level of media interest. Peace support operations, in particular, are subject to intensive media coverage. Unlike conventional war where restrictions are usually imposed on the media, reporters have much more flexibility during peace-support operations to move independently in a country and even across borders. Every action a soldier takes can be broadcast into living rooms in almost real time, and political leaders *believe* they must be prepared to answer for those actions immediately; therefore, the pressure on the political and military leadership to act or to explain an event or incident can be particularly acute.

The media in Canada and in other democracies represent an important conduit to the people.[94] While one author contends that this media focus and associated immediate political reaction will seldom have a strong long-term influence on public opinion, press reporting often has important consequences on decision making:

> Although the media may not have the impact on the substantive policy preferences of the public that some impute to it, technological and other advances could have a profound effect on democratic governance. Perhaps the most important effects would be a perception among policy makers that the electronic media are shortening their decision cycles and the increased availability of "flash" polling that often reflects little more than ephemeral and transitory opinion.[95]

In a democracy, public support for military operations is almost always one of the key centres of gravity of the campaign, and the government will devote significant efforts to maintain high public support for its decisions and actions.[96] The situation gets worse when the media erroneously mis-report or distort the actions taking place for the purpose of undermining government policies, thus sowing doubt in the domestic population.[97] The government, and the military, then has to devote extra efforts to correct inaccuracies, adding pressure to the chain of command at all levels.

The pressure on the military and the politicians created by the CNN effect and the clash between military and media objectives heighten the tension between the military officers "who want to control, as much as possible, everything on the battlefield or area of operations" and reporters who "want unfettered access to all aspects of an operation."[98] With senior political and military leaders held accountable by a media for tactical actions, these leaders will feel compelled to become more engaged and to micromanage the conflict as the situation develops.[99] In Canada the pressure and the dynamics that the daily question period in Parliament generates are, at times, astonishing.[100] This daily event has important repercussions on all levels of command from the strategic to the tactical, imposing high demands to feed the latest information to the politicians and their staff. Thus, it should not be surprising to expect this pressure to create a tendency by the politicians to limit military activities to the more predictable and less risky and to want to control many aspects of the military operation, or to retain many aspects of decision making in Ottawa. As one allied general officer argues, this control may "find its expression in direct interference with the operational and tactical leadership on the ground."[101]

The Legacy of the Somalia Mission and Command Responsibility. In 1995, Minister of National Defence David Collenette established the Somalia Commission of Inquiry to examine all issues relating to the incidents surrounding the deployment of the CF in Somalia during 1992–1993. In the two-year inquiry, the commissioners investigated "the chain of command system, leadership, discipline, and actions and decisions of the Canadian Forces."[102] In its final report released in 1997, the Commission published a separate volume on "the failure of senior [CF] leaders," which identified many aspects of failed leadership and misconduct that were attributed to some military leaders responsible for various components of the mission. In all, eleven officers, including seven general officers, were specifically mentioned for improperly exercising their command responsibility.

The Commission determined that "the failings of the senior leade-rship…can be characterized as inappropriate control and supervision."[103] The CDS of the day was highly criticized for having cultivated an atmosphere that fostered more failings among his subordinates: "…benign neglect, or unquestioning reliance, became so common under General de Chastelain's command that it became every day practice throughout the chain of command."[104] The consequence of this public indictment of many senior officers and commanders continues to have a significant impact on how senior officers approach their command responsibility in 2005.

General and flag officers who are now in the senior command positions of the CF were lieutenant-colonels/commanders and colonels/captains (Navy) at the time of the inquiry, and witnessed the intense scrutiny to which their superiors were subjected; they could also see the negative effect that the Somalia incident and the subsequent inquiry had on the morale of the CF and how they contributed to the gradual loss of public confidence in the CF as an institution. While much of this confidence has been regained with the exemplary performance of CF personnel on many domestic and international operations in the past ten years, such confidence is a fragile commodity that must be constantly renewed through continued exemplary performance. The 1997 *Report to the Prime Minister on the Leadership and Management of the Canadian Forces*, released by Minister Doug Young, mandated many far-reaching changes to improve leadership in the CF. As with the Somalia Inquiry, the implementation of the recommendations of this report also had an important effect on the nature of command in the CF and strongly influenced the way the senior CF leaders regard responsibility and accountability.

Canadian Forces officers now in command take very seriously the doctrine of full responsibility for activities under their command (and the associated accountability that goes with it). However, the enduring effect of the new CF command approach is a culture that does not tolerate errors and that tends to encourage the imposition of additional controls to deal with the risks of command.[105] This is most evident for expeditionary operations, in areas ranging from financial management to medical support to operations planning to mission selection to the use of force. Establishing additional controls throughout the chain of command is often the most convenient way to attempt to create a risk-free command environment, especially when the operational commander and his staff are assembled at the last minute and may not understand the strategic intent in the same way as those who have planned the mission.

Events like those that took place in late 2003 at the American-run Abu Ghraib detention facility in Iraq where US military personnel abused prisoners serve to underline in the mind of some that the establishment of more controls is still the best approach to risk-free command. The civilian-military panel mandated to review the detention operations in Iraq reinforced this view when it concluded "that commanding officers and their staff at various levels failed in their duties and that such failures contributed directly or indirectly to detainee abuse. Commanders are responsible for all their units do or fail to do, and should be held accountable for their action or inaction."[106] The panel also found that the "unclear command structure" was a key contribution to the atmosphere at Abu Ghraib that allowed the abuses to take place.[107] The doctrine of responsibility in the military remains very central to command, as it should be, whether or not the commander knows that incidents are taking place under his watch.[108]

So while it is temping to blame technology, politicians or the media for the compression of the levels of war and the resulting culture that supports centralized decision making, the military senior leadership actually plays a significant part in promoting — or discouraging — this type of culture. As two experts on decision making point out, "evidence shows that when something unexpected happens, this is an unpleasant experience.... A surprise tends to be unpleasant because your world seems to be less predictable and less controllable than you first thought."[109] Adopting mission command in the CF would mean that strategic leaders would have to accept more risks and more uncertainty. However, the acceptance by strategic leaders of uncertainty during military operations may portray them as "incompetent" because they might not have every detail of an operation under their control, in the same fashion that US senior commanders were found to have failed in the performance of their duties at the Abu Ghraib detention facility because "they should have known." Consequently changes to the command culture will not happen overnight in the CF, and a new mission-centric and more risk-taking culture will need to be developed in junior officers so that by the time they reach senior ranks, they are more comfortable taking the right decisions in this ambiguous and complex environment.

The Influence of Domestic and International Law. One final element that contributes significantly to elevating decisions to the strategic level is the growing importance of law, especially international law, in many areas relating to the use of the military in international operations.[110] International law is also the primary legal basis for establishing the mandate for international operations from which the authority to use force is derived; it "provides stability in international relations and an expectation that certain acts or omissions will bring about predictable consequences. Therefore, nations normally comply with international law because it is in their best interest to do so."[111] When military force is used in an armed conflict, the Law of Armed Conflict (LOAC), a constituent of international law, is extremely influential in many areas affecting the conduct of military operations.[112]

The unique characteristics of the type of conflicts that will dominate the future and that Canada intends to participate in, when combined with the demands of the international media and the presence on the ground of influential international non-governmental organizations such as Amnesty International and Human Rights Watch, will continue to result in unprecedented scrutiny of any military action. As a result, ethical and legal considerations will remain much more at the forefront in future conflicts than they have in previous history.[113]

The LOAC has also evolved significantly in the past decade, and the recent creation of the International Criminal Court serves to demonstrate that international law is becoming more dominant in all aspects of military operations.[114] The legal considerations that go into the planning of a specific mission and the number of diverse situations requiring legal counsel that military commanders face during the conduct of military operations cannot be overstated. With "the military lawyer…becoming one of the commander's most important advisors,"[115] there is little decided in international operations without first consulting military lawyers.

With Canada and the CF committed in the future to participate in operations centred on the three-block war scenario — where chaos and complexity are expected to be more prevalent — it is reasonable to expect the legal issues to be more, not less, convoluted. The reality nowadays is that the broad range of international and domestic law expertise required to address the complex issues arising from expeditionary operations seriously limits the authority and freedom of action of deployed Canadian operational commanders. Since they are normally assisted by only one or two relatively junior military lawyers on their national command staffs, most legal issues end up being discussed and resolved in Ottawa where the unique legal expertise to resolve these issues often resides. Two obvious consequences flow from this fact. First, this reach-back invariably means that strategic commanders become engaged in issues that could otherwise be analyzed and decided in theatre, adding to the centralization trend. Second, decision cycles are slowed significantly, especially when the missions area is located several time zones away, potentially resulting in missed opportunities for Canadian commanders in accepting time-sensitive military tasks within a coalition framework.

This need not be the case, however. Canadian operational commanders should be given 24-hour access to specialized legal staff, while retaining the authority for decisions at their level (without engaging any strategic decision makers). Again, a significant change of culture will need to take place for this to happen.

Conclusions

> Militaries claim a desire for mission-oriented command and control, but is this goal achievable? Western societies have developed very centralized organizational structures that promise high level of control. In some ways, such structures are politically and military attractive. But can our cultures tolerate decentralized C^2 philosophies that encourage independent thought and action?
>
> Lieutenant-General Mike K. Jeffery
> Commander of the Canadian Army, 2001–2004[116]

Decision making in the CF for expeditionary operations has changed significantly since the end of the Cold War. The changes have been most noticeable in the past six years, since the Kosovo air campaign of 1999, and have revealed a clear trend toward centralized decision making at the strategic level. The inclination to continue to develop highly centralized organizational structures is going to remain strong, putting more pressure on Canadian C^2 structures for expeditionary operations.

Many forces are contributing to this centralization, and they are not about to go away. The most significant force centres on the increased compression of the levels of war and the emergence of a new paradigm of decision making in military operations. The classical definition of three levels of war is losing its relevance to modern military operations. Issues that used to be handled at one level now cut across all three levels, and decisions that were considered to be the prerogative of commanders at the tactical and operational levels have been raised to the strategic level. Senior commanders contend that the types of issues arising out of modern conflict require they be more actively involved in the day-to-day decision making of operational and tactical issues.

Most military analysts blame the exponential growth of information networks and high-speed global communications for giving politicians and military leaders the tools they yearned to possess to take decisions. Indeed, to date, the greater reliance placed on investing in technology to improve command systems (as opposed to the human element in command) has generally resulted in an additional incentive for senior leaders to pursue the centralization of decision making. As strategic commanders continue to justify their need to be involved in operational and tactical matters for a variety of reasons (including the requirement to provide answers to the politicians), they will continue to build and develop C^2 systems that meet this need.

In Canada many other elements will continue to put pressure on DND to centralize military decision making, and they centre on the unique relationship between the CF, the government and the public. The nature of operations the CF is expected to carry out in future years is such that the degree of autonomy exercised by the Canadian military will likely remain diminished, as compared to conventional war fighting or to other eras when militaries might have been able to operate relatively independently. With the government advocating a more "integrated" approach to guide future Canadian contributions on the international scene in order to achieve greater strategic effects, more stakeholders will partake in national decision making, thereby increasing the central co-ordination required and further reducing the degree of military autonomy. The need to consult legal advisors on complex matters of both domestic and international law will likely continue to drive operational commanders to raise issues to the strategic headquarters in Ottawa, and this will slow down decision cycles. However, the strong influence of the Canadian national media and the unique nature of Canada's parliamentary democracy will continue to put pressures on politicians and military leaders for faster decision cycles. The experiences of the Kosovo air campaign and the operations in Afghanistan and the Persian

Gulf in the past four years certainly point to continued interest and involvement by politicians in military matters.

Finally, in the years since the Somalia Inquiry a risk-averse culture has developed in the CF, one that is not about to be reversed easily. While it is certainly welcoming and highly refreshing for the new CDS to speak of instituting a mission-centric approach to command in the CF, events such as those in Somalia in the 1990s and more recently at the Abu Ghraib detention facility only fuel the demands for increased centralization and more controls, not less. As Lieutenant-General Jeffery queried, "can our cultures tolerate decentralized C2 philosophies that encourage independent thought and action?"[117] Changing the risk-averse approach that exists in the CF is certainly one of the best means to adopt the principles behind mission command. Commanders at all levels will have to be comfortable in dealing with more uncertainty, in tolerating a greater dosage of chaos than expected and, most significantly, in being able to cope with the risks that lower-level commanders are prepared to assume based on their delegated level of authority. This is, in the end, truly the crux of the issue in adopting the mission-command philosophy of leadership in modern conflict. Much work will have to be done in this regard to change the Canadian military and political cultures.

This chapter has argued that mission command has for all intents and purposes disappeared as a command philosophy in the CF. The impact of the loss is even more significant at the strategic-operational level boundary for expeditionary operations since this interface is the critical node for effective national command. For a number of reasons, a new philosophy of command has evolved in the past fifteen years in the CF, and it has significantly changed the role of the strategic, operational and tactical commanders. If the centralization trend continues, the role of the deployed Canadian operational commander, as conceptually envisaged in doctrine, should be revisited.

Clearly, adopting Moltke's mission-command philosophy is too simplistic a solution to address today's C^2 challenges. As Pigeau and McCann have said, the "point is not to argue that micro-management is always wrong or that mission command is always right — that would trivialize the complexity of military operations"; rather, it is important to find "the correct balance between encouraging creative command and controlling command creativity."[118] Consequently, in devising a CF command philosophy for tomorrow, the challenge for the CF is to create a Canadian command framework developed for Canadian national requirements, using contemporary command principles, and with the management of risk as the key factor influencing the development of control structures and processes. In short, new command protocols need to be established, consistent with the authority delegated to operational and tactical commanders. Aligning both *command* and *control* is key to progressing towards General Hillier's vision of a more mission-command philosophy.

NOTES

1   General R.J. Hillier, opening address to the Standing Committee on National Defence and Veterans Affairs, 19 May 2005.

2   Record of decisions from the 18 April 2005 Armed Forces Council. Used with permission of the Director, National Defence Headquarters Secretariat.

3   United Kingdom, Vice-Chief of the Defence Staff, *The UK Joint High Level Operational Concept* (March 2005), 4–1.

4   An example of the many restrictions and regulations can be found in the Deputy Chief of the Defence Staff directive to designated commander for international operations, the "DCDS Directive for International Operations," a document that was barely 50 pages when this author participated in a United Nations mission in 1994 and that has now reached 494 pages.

5   This assertion is also based on the personal experience of the author as Chief of Staff for Joint Task Force South-West Asia (Operation Apollo) and Commanding Officer of the National Command Element, from discussions with recent task force commanders on other operations, and from the Operation Apollo post-operation report, DCDS, "Operation APOLLO Lessons Learned Staff Action Directive," NDHQ, 3350-165/A27, dated April 2003, B-15.

6   References abound on this issue. The following four provide a glimpse of the changes since the mid-1990s. Thomas J. Czerwinski, "Command and Control at the Crossroads," *Parameters* 26, no. 3 (Autumn 1996), 121–32; Wesley Clark, *Modern War: Bosnia, Kosovo and the Future of Conflict*, 2nd ed. (New York: Public Affairs, 2001), 85–6; Christopher J. Bowie, Robert P. Haffa, Jr., and Robert E. Mullins, "Trends in Future Warfare," *Joint Force Quarterly* no. 35, (nd), 129–33; and in Canada, Richard Goette, "Command and Control Implications for Canadian Forces Air Expeditionary Operations," in *Canadian Expeditionary Air Forces*, ed. Allan D. English (Winnipeg: Centre for Defence and Security Studies, 2002), 67–82.

7   See, for instance, the recently released *The UK Joint High Level Operational Concept*, 4–1. Also, in the US, see Milan Vego, "What Can We Learn from Enduring Freedom?" *US Naval Institute Proceedings* 128, no. 7 (July 2002), 28–33.

8   Vego, "What Can We Learn from Enduring Freedom?" 30–1. For the same argument, documented during the Kosovo campaign, see Benjamin S. Lambeth, "Lessons from the War in Kosovo," *Joint Force Quarterly* no. 30 (Spring 2002), 14.

9   See for instance Douglas A. Macgregor, "Future Battle: The Merging Levels of War," *Parameters* 22, no. 4, 33–47; and Martin Dunn, "Levels of War: Just a Set of Labels?" *Australian Army's On Line Journal*, Research and Analysis: Newsletter of the Directorate of Army Research and Analysis, no. 10, Canberra (October 1996), 1–4.

10   William A. Owens, *Lifting the Fog of War* (Baltimore: John Hopkins University Press, 2000), 23.

11   See notably Wesley Clark, *Modern War*, xxxviii–xxxix, 85–6; Thomas-Durell Young, "NATO

Command and Control for the 21st Century," *Joint Force Quarterly* no. 29 (Autumn/Winter 2000–2001), 40–5; Mark G. Davis, "Centralized Control/Decentralized Execution in the Era of Forward Reach," *Joint Force Quarterly* 35 (nd), 95–9; and Milan N. Vego, "Operational Command and Control in the Information Age," *Joint Force Quarterly* 35 (nd), 100–107, for discussions on this theme; and in Canada, see Allan English, "Rethinking 'Centralized Command and Decentralized Execution,'" in *Air Force Command and Control*, eds. Douglas L. Erlandson and Allan English (Toronto: Canadian Forces College, 2002), 71–81.

12   Glossary in DND, *Leadership in the Canadian Forces: Conceptual Foundation* (Kingston: Canadian Defence Academy, 2005), 131.

13   *UK Joint High Level Operational Concept*, Foreword and p. 4-1.

14   DND, *Leadership in the Canadian Forces: Doctrine* (Kingston: Canadian Defence Academy, 2005), 12–13.

15   Czerwinski, "Command and Control at the Crossroads," 122.

16   Van Creveld, *Command in War*, 10.

17   See Michael DeLong, *Inside CentCom: The Unvarnished Truth About the War in Afghanistan and Iraq* (Washington, DC: Regnery Publishing, 2004), 26; and Vego, "What Can We Learn from Enduring Freedom?" One analysis concluded that this "command from Tampa" was not effective: "a number of command-and-control decisions Franks made leading up to [Operation] Anaconda [in Afghanistan], as well as poor communications between various components commanders, set the stage for confusion and questionable battle-planning decisions." Major Mark Davis, MA thesis, School of Advanced Air and Space Studies, Maxwell AFB, as quoted by Elaine Grossman, "Inside the Pentagon," 29 July 2004, 1.

18   Van Creveld, *Command in War*, 53.

19   Czerwinski, "Command and Control at the Crossroads," 123.

20   Van Creveld, *Command in War*, 144.

21   See Major-General Werner Widder, "Auftragstaktik and Innere Führung: Trademarks of German Leadership," *Military Review* 82, no. 5 (September–October 2002), 3–9. The term *task-oriented command* is also used at times. For simplicity and consistency, *mission command* will be used throughout.

22   Moltke, as quoted in Widder, "Auftragstaktik and Innere Führung," 4.

23   Ibid.

24   Van Creveld, *Command in War*, 188.

25   Czerwinski, "Command and Control at the Crossroads," 124–5.

26   Van Creveld, *Command in War*, 270.

27   Ibid., 268.

28   John Keegan, as quoted in David S. Alberts and Richard E. Hayes, *Power to the Edge: Command and Control in the Information Age* (Washington, DC: CCRP Publication Series, 2004), 48.

29   Van Creveld, *Command in War*, 274.

30   Christopher D. Kolenda, "Transforming How We Fight: A Conceptual Approach," *Naval War College Review* 56 (Spring 2003), 110.

31   *Network-centric warfare* (NCW) is defined as the notion that the information superiority–enabled concepts of operations generate increased combat power. This is achieved by networking sensors, decision makers, and shooters to achieve shared awareness, increased speed of command, higher tempo of operations, greater lethality, increased survivability, and a degree of self-synchronization. David S. Alberts, John J. Gartska and Frederick P. Stein, *Network Centric Warfare: Developing and Leveraging Information Superiority* (Washington, DC: National Defense University Press, 2000), 2.

32   Ross Pigeau and Carol McCann, "Re-conceptualizing Command and Control," *Canadian Military Journal* 3, no. 1 (Spring 2002), 53–63.

33   Ibid., 56. The CF and NATO definition of *command* is "the authority vested in an individual of the armed forces for the direction, co-ordination, and control of military forces." DND, *Canadian Forces Operations*, B-GG-005-004/AF-000, p. 2-1.

34   Pigeau and McCann, "Re-conceptualizing Command and Control," 57.

35   Ibid., 62.

36   Vego, "What Can We Learn from Enduring Freedom?" Further, he recommends that "the German-style, task-oriented command and control from top to bottom [be] adopted. Otherwise the [US] Armed Forces could find themselves resembling the former Soviet military and paying a heavy price in the quest for absolute certainty and control."

37   For instance, Colonel Christian Rousseau contends that Western militaries have been slow to recognize that a mission-command approach is what is required to deal with complexity; moreover, he adds that "even now, whenever technology floats the mirage of complete visibility of the battlespace, we let ourselves be tempted by the allure of more control. Unless the complete visibility promised also includes complete structural information (and it cannot), mission command remains the only viable alternative." Christian Rousseau, "Complexity and the Limits of Modern Battlespace Visualization," *Canadian Military Journal* 4, no. 1 (Spring 2003), 42. See also Rousseau's chapter in this book.

38   Pigeau and McCann, "Re-conceptualizing Command and Control," 57.

39   Emphasis added. Lieutenant-General R.R. Henault, "Modern Canadian Generalship in Conflict Resolution: Kosovo as a Case Study," in *Generalship and the Art of the Admiral*, eds. Bernd Horn and Stephen Harris (Toronto: Vanwell Publishing, 2000), 288.

40   A plethora of joint-doctrine publications have been produced, providing guidance and direction to the joint staff, commands, formations and agencies supporting operations. For force employment, the key manuals include *Canadian Forces Operations* (often referred to as the *CF Operations* manual), *Use of Force in CF Operations* (often referred to as the *Use of Force* manual), *Risk Management for CF Operations, and DCDS Directive for International Operations* (DDIOs). See the DCDS Group Joint CF Doctrine Web page for copies of these manuals at http://www.dcds.forces.gc.ca/jointDoc/ pages/J7doc_keydocs_e.asp.

41   Joint task forces are constituted when two or more environmental elements of the CF participate in the operation. When there is only one element, it is referred to as a task force. In such operations the DCDS has the responsibility to co-ordinate, on behalf of the CDS, strategic-level operational planning with and operational direction to the task force commander.

42   DND, *Canadian Forces Operations*, p. 7-2. *Operational commander* usually means the commander conducting operations, while *operational-level commander* is a commander employed in a formation or headquarters at the operational level. *Canadian Forces Operations*, p. 8-2.

43   Henault, "Modern Canadian Generalship in Conflict Resolution," 287.

44   This situation is not to be confused with the recent experience of General Hillier who, as Commander of the International Security Assistance Force filling a senior NATO alliance command position, was genuinely *functioning* at the operational level. An NCE, commanded by a Canadian colonel, was also established in Kabul to deal with Canadian-unique matters (essentially liaison, administration and logistics).

45   Henault, "Modern Canadian Generalship in Conflict Resolution," 288.

46   Ibid.; emphasis added.

47   Douglas Bland, *Chiefs of Defence* (Toronto: Canadian Institute of Strategic Studies, 1995), 198.

48   See the Canadian keystone joint-doctrine manual, *Canadian Forces Operations*, 1–4.

49   Clark, *Waging Modern War*, 86.

50   Widder, "Auftragstaktik and Innere Führung," 6.

51   Vego, "Operational Command and Control in the Information Age," 101–102.

52   Ibid., 102.

53   David C. Gompert et al., "Stretching the Network," RAND Corporation Occasional Paper, April 2004.

54   Vego, "Operational Command and Control in the Information Age," 103.

55   Ibid., 102.

56   DeLong, *Inside CentCom*, 26.

57   Owens, *Lifting the Fog of War*, 141–42.

58   Ibid., 205.

59   Elinor Sloan, *The Revolution in Military Affairs* (Kingston: McGill-Queen's University Press, 2002), 15.

60   A.R. Cebrowski, "The Implementation of Network-Centric Warfare," U.S. Department of Defense Office of Force Transformation, (January 2005), 9–10.

61   Sloan, *The Revolution in Military Affairs*, 15.

62   General Krulak, a former US Marine Corps commandant, first depicted this new phenomenon in a 1999 article on the "three-block war," highlighting the new kind of soldiers that would be needed for future conflicts. See Charles Krulak, "The Strategic Corporal: Leadership in the Three-Block War," *Marine Corps Gazette* 83, no. 1 (January 1999), 18–22.

63   As one former division commander with extensive peacekeeping experience explained, during peace-support missions the soldier's role is even more critical: "It is frequently important to uphold the principle of impartiality, especially under difficult circumstances. In this environment, the still-smoldering fuse of the power keg can be quickly reignited, and the peace force can become the enemy of one faction or another. Such a loss of credibility would have serious political implications." Widder, "Auftragstaktik and Innere Führung," 6.

64   Australian Army, Future Land Warfare Branch, "Complex Warfighting," concept paper (April 2004), 8. As this study indicates, "the operational level of war may be disappearing, 'squeezed out' by direct interaction of tactical actions with strategic outcomes," adding to the compression of the levels of war.

65   See Sean M. Maloney and Douglas L. Bland, *Campaigns for International Security* (Kingston: McGill-Queen's University Press, 2004), for a description of operations of the past fifteen years.

66   Canada, *A Role of Pride and Influence in the World: Defence*, Canada's International Policy Statement (April 2005), 8–11.

67   Martin L. Cook, "The Proper Role of Professional Military Advice in Contemporary Use of Force," *Parameters* 32, no. 4 (Winter 2002–2003), 30.

68   For a more detailed discussion on the challenges of coalition warfare see Robert W. Riscassi, "Principles for Coalition Warfare," *Joint Forces Quarterly* no. 1 (Summer 1993), 58–71. For a Canadian perspective see Douglas Bland, "Canada and Military Coalitions: Where, How and with Whom?" *Policy Matters* 3, no. 3 (Montreal: Institute for Research on Public Policy, February 2002).

69   Henault, "Modern Canadian Generalship in Conflict Resolution," 289.

70   See Lambeth, "Lessons from the War in Kosovo," 13. For a discussion on Canadian national command, see Ross Graham, "Civil Control of the Canadian Forces: National Direction and National Command," *Canadian Military Journal* 3, no. 1 (Spring 2002), 23–29.

71   Mike Greenley et al., "Complex Situations: DND Joint Command and Decision Making in Response to Asymmetric Threats, Terrorism and other Complex Operational Scenarios," Defence Research Development Canada, Valcartier, Report CR2004-272 (November 2004), 21–6. Table 3 of the report offers a complete and comprehensive list of the variables for a peace-support operation. The report is focused on offering a list of possible mitigation strategies — largely centred on creating more tools at all levels of command and establishing information technology systems — to reduce the impact of any given variable and to therefore decrease complexity.

72   Canada, *A Role of Pride and Influence in the World: Defence*, 8–10.

73   The co-ordination is most often done at the Privy Council Office and will involve representatives from Defence, Foreign Affairs Canada and the Canadian International Development Agency. As used here, operational level is meant to include the theatre level of war, a level often recently referred to in many publications.

74   Canada, *A Role of Pride and Influence in the World: Defence*, 6.

75   For a more complete discussion on this theme see, notably, Gary A. Klein, "Why Good People Make Bad Decisions," *Sources of Power: How People Make Decisions* (Cambridge, MA: MIT Press, 1998), 271–85; and Dietrich Dörner, *The Logic of Failure: Recognizing and Avoiding Error in Complex Situations* (New York: Metropolitan Books, 1996).

76   See Elliot A. Cohen, *Supreme Command: Soldiers, Statesmen and Leadership in Wartime* (New York: Anchor Books, 2002).

77   Lloyd J. Matthews, "The Politician as Operational Commander," *Army* 46, no. 3 (March 1996), 30.

78   Cook, "The Proper Role of Professional Military Advice," 30.

79   Samuel P. Huntington, *The Soldier and the State* (Cambridge, MA: Belknap Press, 1957; 12th printing, 2003) 80–5.

80   Civil control of the military in most Western democracies is managed and maintained through a sharing of responsibilities between civilian leaders and military officers. For a discussion on this issue see Douglas Bland, "A Unified Theory of Civil-Military Relations," *Armed Forces and Society* 26 (Fall 1999), 7–26.

81   For an example of a controversial mission, consider the use of Joint Task Force 2 for special operations outside Canada. For sensitivities with targeting, see Frederic L. Borch, "Targeting After Kosovo: Has the Law Changed for Strike Planners," *Naval War College Review* 56 (Spring 2003), 66–8.

82   See notably Douglas Bland, "Military Command in Canada," in *Generalship and the Art of the Admiral*, 121–36; and Shamir and Ben-Ari, "Challenges of Military Leadership in Changing Armies," *Journal of Political and Military Sociology* 28 (Summer 2000), 43–59.

83   Henault, "Modern Canadian Generalship in Conflict Resolution," 289.

84   Mackenzie King, as quoted in John Macfarlane, *Ernest Lapointe and Quebec's Influence on Canada's Foreign Policy* (Toronto: University of Toronto Press, 1999), 181.

85   Desmond Morton, "The Political Skills of a Canadian General Officer Corps," in *Generalship and the Art of the Admiral*, 370.

86   In the US during the 1962 Cuban missile crisis, the confrontation between Secretary of Defense McNamara and Admiral Anderson, Chief of Naval Operations, over the political nuances stemming from the execution of a naval blockage of Soviet ships, is probably the best-known and most dramatic example of this situation. President Kennedy was worried that the U.S. Navy, "already restive over the controls imposed on how the blockage was to be executed, might 'blunder into an incident'" and the possibility of nuclear war. See Matthews, "The Politician as Operational Commander," 33–5. Canada had its own civil-military crisis at the time of the Cuban missile crisis. Peter T. Haydon, *The 1962 Cuban Missile Crisis: Canadian Involvement Reconsidered* (Toronto: Canadian Institute of Strategic Studies, 1993). Conscription crises in Canada are other examples.

87   Matthews, "The Politician as Operational Commander," 32.

88   Christopher J. Bowie et al., "Trends in Future Warfare," 131.

89   The most popular recent exposés in this regard have been Andrew Cohen, *While Canada Slept* (Toronto: McClelland & Stewart, 2003), and J.L. Granatstein, *Who Killed the Canadian Military?* (Toronto: HarperCollins Publishers, 2004).

90   Lambeth, "Lessons from the War in Kosovo," 15.

91   Bland, "Military Command in Canada," in *Generalship and the Art of the Admiral*, 128.

92   The *CNN effect* is an expression employed to "represent the collective impact of all real-time news coverage." Margaret H. Belknap, "The CNN Effect: Strategic Enabler or Operational Risk?" *Parameters* 32, no. 3 (Autumn 2002), 100.

93   Henault, "Modern Canadian Generalship in Conflict Resolution," 276.

94   See a strong criticism of Canadian generals in Kabul and their relationship with the media in Stephen Thorne, "The Enemy Within," *Canadian War Correspondent Association* (Fall 2004), 17–20.

95   Eric V. Larson, quoted in William M. Darley, "War Policy, Public Support and the Media," *Parameters* 35, no. 2 (Summer 2005), 132.

96   Ralph Coleman, "The General/Admiral's Role in Public Affairs in International Operational Theatres," in *Generalship and the Art of the Admiral*, 384.

97   Darley, "War Policy, Public Support and the Media," 132.

98   Belknap, "The CNN Effect," 108.

99   Ibid., 109.

100   For a critical analysis of Parliament, see Roy Rempel, *The Chatter Box* (Toronto: Breakout Educational Network, 2002).

101   Widder, "Auftragstaktik and Innere Führung," 6.

102   The Commission was also mandated to investigate "the actions and decisions of the Department of National Defence in respect of the Canadian Forces' participation in the peace enforcement mission in Somalia." The Commission of Inquiry into the Deployment of Canadian Forces to Somalia was established on 20 March 1995 under the federal Inquiries Act. Terms of reference of the inquiry appear under Order-in-Council 1995-442 and are enclosed in Commission of Inquiry into the Deployment of Canadian Forces to Somalia, *Dishonoured Legacy: The Lessons of the Somalia Affair*, 5 vols. (Hull: Communications Group, 1997), vol. 5, appendix 1.

103   *Dishonoured Legacy*, vol. 4, 956.

104   Ibid., 956–7.

105   Ironically, Canadian senior officers have written little about this issue in that period. For a recent discussion see Fred Bigelow, "Nothing Ventured, Nothing Gained: Risk Aversion and Command," paper prepared for the Advanced Military Studies Course 7, Canadian Forces College, Toronto (December 2004).

106   James R. Schlesinger et al., *Final Report of the Independent Panel to Review DOD Detention Operations* (August 2004), 43.

107   Ibid., 45.

108   See Elliot A. Cohen and John Gooch, *Military Misfortunes: The Anatomy of Failure in War* (New York: Anchor Books, 1991), chap. 2, "Understanding Disaster," and chap. 3, "Analyzing Failure," 5–57.

109   Karl E. Weick and Kathleen M. Sutcliffe, *Managing the Unexpected: Assuring High Performance in an Age of Complexity* (San Francisco, CA: Jossey-Bass, 2001), 39.

110   International law covers many areas from the control of the use of force to maritime navigation in high seas to civil aviation to the treatment of combatants and civilians on the battlefield.

111   DND, *Use of Force in CF Operations*, B-GJ-005-501/FP-000 (1 June 2001), 1–2.

112   The Law of Armed Conflict "arises from a desire among civilized nations to prevent unnecessary suffering and destruction while not impeding the effective waging of war. A part of public international law, LOAC regulates the conduct of armed hostilities. It also aims to protect civilians, prisoners of war, the wounded, sick, and shipwrecked. LOAC applies to international armed conflicts and in the conduct of military operations and related activities in armed conflict." From the U.S. Military Web site, http://usmilitary.about.com/cs/wars/a/loac.htm.

113   Henault, "Modern Canadian Generalship in Conflict Resolution," 289.

114   See Borch, "Targeting After Kosovo: Has the Law Changed for Strike Planners?" for a discussion on targeting and the LOAC.

115   Henault, "Modern Canadian Generalship in Conflict Resolution," 285.

116   M.K. Jeffery, "Foreword" to *The Human in Command: Exploring the Modern Military Experience* by Carol McCann and Ross Pigeau (New York: Kluwer Academic / Plenum Publishers, 2000), v.

117   Ibid.

118   Pigeau and McCann, "Re-conceptualizing Command and Control," 57.

CONCLUSION

Allan English

The purpose of this book was to offer Canadian perspectives on arguably the most important aspects of operational art: leadership and command. The perspectives are meant to complement those found in a previous book, *The Operational Art: Canadian Perspectives — Context and Concepts*, in the series. Both books share the premise that there are unique Canadian approaches to operational art based on our national and military culture and historical experience. Canadian military professionals should become familiar with these approaches so that the practice of their profession will be based on sound theory as well as their personal experience. As noted in the introduction to this book and in the CF profession of arms manual, *Duty with Honour*, professional expertise rests on the mastery of relevant theoretical knowledge as well as practical skills. Most of the contributors to this volume bring insights into leadership and command and the operational art based on both practical experience and rigorous academic study.

The Operational Art: Canadian Perspectives — Leadership and Command began by affirming a truth that has sometimes become lost at the beginning of the twenty-first century as slogans like "everything is joint now" abound and obscure the reality that the physical and cultural settings in which armed forces operate shape their leadership and command styles. Much of the literature on leadership and command, as well as joint doctrine and operational art, especially in Canada, is dominated by land force or army experiences. Now that we are beginning to document Canadian Navy approaches to leadership and command, it is to be hoped that the Canadian Air Force approaches will be similarly documented so we may have a more balanced approach to leadership and command in this country.

While there are many similarities among Environmental (or service) leadership and command styles, the fact remains that there are also significant differences and that a "one size fits all" approach to leadership and command will not work in many circumstances. The challenge for practitioners of the operational art is to recognize the appropriate circumstances in which to use joint command and leadership frameworks to co-ordinate the actions of land, air and maritime forces, whether it be in a domestic or multinational setting. Another challenge for practitioners of the operational art is to recognize when different (some use the term *integrated* today) command and leadership frameworks will be required to co-ordinate the activities of government agencies and non-government agencies with those of military forces.

The concept of *common intent*, examined by Ross Pigeau and Carol McCann in this book, is one that can apply in many different circumstances and at all levels of operations, from the tactical to the strategic, and Pigeau and McCann offer a number of concrete ways to use this theoretical tool in practical situations. Establishing common intent can be one of the greatest challenges in joint and integrated operations where the differences in national and organizational culture are frequently barriers to its creation. The experiences of the Canadian Army's stabilization efforts in post-conflict Afghanistan and the Canadian Navy's command of coalition operations in the Arabian Sea are examples of the great success Canadians have had in establishing common intent in human networks composed of diverse cultures.

Despite these successes, the evolution of CF joint command and control structures has been ad hoc and often rushed. Without an effective method of collecting and disseminating lessons learned from various post-Cold War CF operations and with inadequate CF joint doctrine, the CF depended on a group of senior leaders and experienced staff officers to cobble together structures while conducting operations. The CF was fortunate to have leaders of the calibre of General Henault and to have a cadre of competent staff officers who worked together as a team while these systems were evolving. The expertise and the continuity they provided overcame many of the obstacles they faced. However, few, if any, of those involved in this process would advocate using it in the future. As General Henault noted, a proper lessons-learned process, effective doctrine and, I would add, relevant theory are required to ensure the successful evolution of CF command and control structures to meet future challenges. This should come as no surprise because the ideal doctrine cycle consists of all of these components, as shown in Figure 10.1.

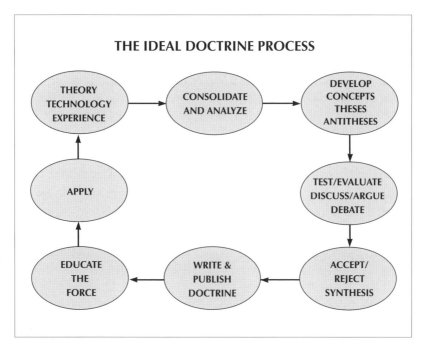

FIGURE 10.1. THE IDEAL DOCTRINE PROCESS[1]

Lacking comprehensive analyses of Canadian experience, it would appear that Canadian doctrine and theory on the operational art, including that on leadership and command, is largely lacking, and what exists is deficient in many ways. This deficiency is having an impact on current CF transformation efforts, as Daniel Gosselin has argued in his chapter on the continued relevance of mission command and Canadian operational commanders to Canadian expeditionary operations. The deficiency has also had an impact on Canadian professional military education, which is often dependent on foreign doctrine or on "cut and paste" Canadian doctrine based on foreign doctrine.[2] Since theory, experience and doctrine are all interrelated, each informing the other, in the future the Canadian Forces will need to work at improving the analysis and dissemination of its experience, the development of its own theories, the writing of its doctrine, and the integration of all of these activities if it is to optimize its leadership and command in the practice of the operational art. This book is offered as a contribution to that work.

NOTES

1   From Dennis M. Drew, "Inventing a Doctrine Process," *Airpower Journal* 9, no. 4 (Winter 1995), 42–52.

2   For example, the Canadian Forces College (CFC) made this statement about the doctrine being used to teach the Advanced Military Studies Course (AMSC), a senior officer course focusing on the operational art: "Note: As of this year, CAS [Chief of the Air Staff] has discarded *Out of the Sun* as restrictive and inadequate. New doctrine is to be drafted in the coming years by the new Air Warfare Centre. In lieu of Canadian-sanctioned doctrine, CFC will rely on USAF and US DOD Joint Air doctrine." AMSC Schedule for 27 September 2005, "A/JC/CPT 404/LE-3, Nature of Air Operations," accessed 15 October 2005.

CONTRIBUTORS

HOWARD G. COOMBS, BA (Hon), BEd, Master of Military Art and Science (Theatre Operations), retired from full time service with the Canadian Forces in 2002. He is a graduate of the Canadian Forces Staff School, Canadian Land Force Command and Staff College, United States Army Command and General Staff College, where he was one of eleven students who earned the designation US Army Master Strategist in 2001, and the US Army School of Advanced Military Studies, which awarded his Master's degree. He is currently a doctoral candidate at Queen's University in Kingston studying twentieth century Canadian military history. He is also a Teaching Fellow at Queen's, a Research Associate of the Canadian Forces Leadership Institute, Kingston, a part-time instructor at the Canadian Forces College, Toronto and a reserve officer commanding the Princess of Wales' Own Regiment, an infantry unit based in Kingston.

DR ALLAN ENGLISH was the lead academic for the Advanced Military Studies Course from its inception in 1998 to 2004 and he was Co-chair of the Aerospace Studies Department from 2001 to 2005 at the Canadian Forces College. He is an Adjunct Associate Professor of History at Queen's University where he teaches a graduate course in Canadian military history. His book, *Understanding Military Culture: A Canadian Perspective,* is published by McGill-Queen's University Press. In the fall of 2005 the Canadian Defence Academy Press published *The Operational Art: Canadian Perspectives — Context and Concepts* edited by Dr English, Brigadier-General Daniel Gosselin, Howard Coombs, and Captain (Navy) Laurence M. Hickey.

DR RICHARD GIMBLETT served for 27 years in the Canadian Navy prior to becoming an independent historian and defence policy analyst. He served in ships of various classes on both coasts, including as Combat Officer of HMCS *Protecteur* for operations in the Persian Gulf during the war of 1991. His last appointment was to the Directorate of Maritime Strategy, as lead writer of *Leadmark: The Navy's Strategy for 2020.* His newest book, published in June 2004, is *Operation Apollo: The Golden Age of the Canadian Navy in the War Against Terrorism.* He was recently elected President of the Canadian Nautical Research Society. He is a Fellow of the Canadian Defence and Foreign Affairs Institute, a Research Associate with the Canadian Forces Leadership Institute, and is on the Visiting Faculty of the Canadian Forces College.

MAJOR-GENERAL DANIEL GOSSELIN is the Director General – International Security Policy at National Defence Headquarters. Some of his recent assignments include the Chief of Staff of the CF Transformation Team, Commandant of the Canadian Forces College, Special Assistant to the Deputy Chief of the Defence Staff, Commander of the CF Joint Operations Group, and Chief of Staff for the Headquarters, Joint Task Force South-West Asia during Operation APOLLO. He is a graduate of the Advanced Military Studies Course and the National Security Studies Course. He holds a BASc, a MASc, a MPA, and a MA in War Studies. He is also currently a doctoral candidate at Queen's University in Kingston.

COMMANDER KEN HANSEN is the Military Co-Chair of the Maritime Studies Program at the Canadian Forces College, Toronto. His other appointments have included Flag-Lieutenant to the Maritime Commander; Career Manager for MARS Officers; Squadron Weapons Officer with the First Destroyer Squadron; and Senior Staff Officer Fleet Replacement. He completed a Master of Arts degree in War Studies from the Royal Military College of Canada in 2005, winning the Barry D. Hunt Memorial Prize as the top graduate student. His thesis, "Fuel Endurance and Replenishment at Sea in the Royal Canadian Navy, 1935-1945," was awarded the Jacques Cartier Prize by the Canadian Nautical Research Society as the year's best graduate thesis on a nautical subject in Canada.

GENERAL RAY HENAULT served in number of command and staff appointments in the air force as a pilot and flight instructor and as an air traffic controller. He is a graduate of the École supérieure de guerre aérienne (ESGA) in Paris and Canada's National Defence College. He has a Bachelor of Arts degree and an Honorary Doctorate of Laws from the University of Manitoba. In 1996 he was posted to Ottawa, beginning a series of appointments at National Defence Headquarters including Deputy Chief of the Defence Staff (DCDS). His three-year tenure as DCDS was highlighted by the Canadian contribution to the Kosovo air and ground campaigns and other significant NATO missions, including the

Stabilization Force in Bosnia-Herzegovina. He was Chief of the Defence Staff from June 2001 until February 2005, a period marked by the highest operational tempo for the Canadian Forces in 50 years. General Henault was elected to the position of Chairman of the NATO Military Committee in November 2004 and assumed that position at NATO Headquarters in Brussels, in June 2005.

GENERAL RICK HILLIER, BSc, is currently the Chief of Defence Staff (CDS) for the Canadian Forces. Commissioned as an armour officer, he served with the 8th Canadian Hussars (Princess Louise's) and later the Royal Canadian Dragoons. Throughout his career, he has commanded men and women from platoon to division and worked as a staff officer in both the Army headquarters in Montreal and the National Defence Headquarters in Ottawa. He has served throughout Canada, twice in both Europe and the United States, as well as with United Nations and North Atlantic Treaty Organization forces in the former Yugoslavia. In 1998 he was appointed the Canadian Deputy Commanding General of III Armoured Corps, US Army, in Fort Hood, Texas and, following that, in 2000, as Commander, Multinational Division (Southwest) in Bosnia. Upon his return to Canada he assumed the duties of Assistant Chief of Land Staff and in 2003 was appointed as the Chief of the Land Staff. During 2004 he commanded the NATO International and Security Assistance Force Rotation V in Afghanistan. He was appointed CDS in February 2005.

CAROL MCCANN heads the Command Effectiveness and Behaviour Section at Defence Research and Development Canada (DRDC)– Toronto. Since receiving an MASc from the University of Toronto in 1979, she has been extensively involved in investigating the human factors aspects of command and control, especially in the development of C^2 concepts, systems and doctrine for the army and navy. She has carried out research in the areas of multimodal human-computer dialogue, decision support, cognitive processes in military planning, human performance under stress, and most recently, team decision making. Together with Ross Pigeau at DRDC Toronto, she has been developing a new theoretical perspective on C^2 that focuses on the human as the most important component. Over the last few years, they have worked together to build an expanded research program on Command Effectiveness and Behaviour emphasizing uniquely human aspects of C^2 such as hardiness, trust, confidence, team decision-making and leadership. Carol has both chaired and participated in NATO Research Study Groups on command and control and currently chairs HUM Technical Panel 11 on Human Aspects of Command.

DR ROSS A. PIGEAU is Chief Scientist for Defence Research and Development Canada in Toronto. He received a BA from Brock University in 1978 and both an MA (1978) and PhD (1985) in experimental psychology from Carleton University. Early in his career, Dr Pigeau conducted research on sleep deprivation and brain electro-physiology as well as studied vigilance fatigue among NORAD surveillance operators. In 1993, he started the Command and Control research program that emphasizes uniquely human aspects of C^2 such as trust, confidence, team decision-making, fatigue and leadership. Together with a Carol McCann he developed a new theory of C^2, one that is influencing military thought both nationally and internationally. He lectured for 17 years on the psychology of command and control at the Maritime Warfare Centre in Halifax, speaks regularly at the Canadian Forces Staff College, contributed to the Canadian Forces profession of arms manual, *Duty With Honour*, and sits on the five-nation Technical Cooperation Panel (TTCP) representing the human sciences.

COLONEL CHRISTIAN ROUSSEAU has commanded at the unit and formation level and has held staff appointments in operational and strategic level headquarters on operations in Canada and abroad. He holds a BSc in science (Mathematics and Physics) from the Royal Military College of Canada. As the current Director General - Military Engineering at National Defence Headquarters, Colonel Rousseau is the Chief Military Engineer of the Canadian Forces.

BRIGADIER-GENERAL (RETIRED) JOE SHARPE joined the Royal Canadian Air Force in 1965 under the Regular Officer Training Plan. He attended Royal Roads Military College in Victoria, BC and graduated from the Royal Military College of Canada in 1969. He served in the Canadian Forces (CF) for the next 32 years in various operational, instructional and staff positions, including Commandant of the CF School of Aerospace Studies and Wing Commander of 17 Wing Winnipeg. He served on the Joint Staff during the Gulf War (1991) and he was the Air Component Commander in the Joint Headquarters during Operation Assistance, the

CF response to flooding in Manitoba in 1997. During his career he attended the Aerospace Systems Course, Command and Staff College and National Defence College. He also spent a year as a senior fellow with the Canadian Institute for International Peace and Security. He chaired the Croatia Board of Inquiry that investigated the medical problems being suffered by CF soldiers returning from peacekeeping operations. He also chaired the Special Review Group commissioned by the CDS to examine issues surrounding CF leadership during the Croatia deployments. He is currently serving as a special advisor to the CF/DND Ombudsman on Post Traumatic Stress Disorder.

COLONEL C.J. WEICKER is a Communications and Electronics officer who has served in a number of field and staff positions. He was the commander of the 1st Canadian Division Headquarters and Signal Regiment and during that time the Regiment was involved in Operation Recuperation, the CF response to the aftermath of the ice storm that had hit parts of Ontario and Quebec at the beginning of 1998, and Operation Abacus, the CF's participation in Canada's Y2K preparations. He was promoted to Colonel on 15 April 2000 and appointed J6 Co-ordination. From May to November 2001 he was deployed overseas as the Chief J6 for the Stabilization Force (SFOR), where he was responsible for the provision of communication and information services for NATO throughout Bosnia and Croatia. He is currently the Commander of the Communication Reserve and the Communications and Electronics Branch Advisor.

INDEX

Critchley, Brigadier General A.C., 8-9

culture 3, 22, 31-2, 34, 74, 174, 218; Canadian military, 1, 2, 7, 35, 48, 76, 195, 217, 218, 220, 222; Canadian naval, 7, 18-19, 43-6, 170; Environmental (service) differences in, 6-7; military, 3, 6-7, 22, 25-7, 74-6, 218; naval, 18-19, 36-7, 39, 41-7; US military, 6-7, 33-4

decision making: in a complex system, 61-76; in modern conflict, 204-20; naturalistic, 63, 65, 77; recognition-primed decision making (RPD), 66-8

DeLong, Lieutenant-General Michael, 206

Effects Based Operations (EBO), 86

emergent behaviour, 87-91

Forcier, Vice-Admiral J.Y., 14

Ghani, Dr. Ashraf, 180

Haiti, 210

Harris, Sir Arthur "Bomber," 14

Hawker, Lanoe, 11

Henault, General Raymond, 135-6, 200, 201, 202, 203, 209, 230; experiences, 141-53; observations on CF command and control, 154-6

Hillier, General Rick (R.J.), 148, 193, 204, 223

integrated operations, x, 1, 3, 4, 25, 148, 153, 158, 221, 230

intent, 73, 74, 75, 90-2, 95-6; commander's, 34, 75, 85, 86, 94, 96-7; common, 86, 90, 91-2, 100, 101-107, 230; explicit, 34, 39, 74-5, 91-2, 95-104; implicit, 34, 39, 74-5, 91-2, 95-104; shared, 39; societal, 85; and will, 100

Operation Recuperation (ice storm), 147, 153

Operation Saguenay, 141-3, 157

operational art, ix , 157

operational planning process (OPP), 64-5, 67, 73-7, 87, 121, 122, 124, 126

peace support operations (PSOs), 173, 175-7, 184-9, 209-210

Pearson, Lester, 174

Persian Gulf, 221-2

Persian Gulf War (Gulf War) (see also Operation Apollo), 47, 109, 135, 137, 140, 141, 142, 149, 150, 167-70, 208

Pratt, David, 176

profession of arms, 6; Canadian, xii-xiv, 3, 25, 26

Professional Military Education (PME), xii-xiv, 77, 231

Revolution in Military Affairs (RMA), 7, 31,165, 173, 186

Roy, Lieutenant-General Armand, 144

Royal Air Force, 8, 9, 10, 14-15; Bomber Command, 11-13

Royal Canadian Air Force (see also Canadian Air Force), 8, 15-16, 22

Royal Canadian Naval Reserve, 18-19

Royal Canadian Naval Voluntary Reserve, 18-19

Royal Canadian Navy (see also Canadian Navy), 18-19, 22, 43-7, 169-70

Royal Flying Corps, 8, 9, 10, 14-15